ARCH
I
TECT

HOW TO BUILD A PYRAMID

By
Osiadan BoreBore Oboadee
November 11, 2011

African Creation Energy
WWW.AfricanCreationEnergy.COM

www.AfricanCreationEnergy.com

ISBN 978-1-105-06816-4

Printed in the United States of America

AFRICAN CREATION ENERGY ALERT:

*** Hat Hard Area ***

*** Min at Work ***

Proceeded if you Will.

"BUILT FROM THE TOP DOWN"

African Creation Energy

Creative Solution-Based Technical Consulting

.:.

Dedicated to

the P.T.A. and the Resurrection

.:. of the Craft of Pyramid Building .:.

 # DEDICATION

netjeru neb redit esen nefer rudj
mer pen kat ten net nefer ka ro
inetisen seped esen inetisen uash esen
inetisen ba esen ineti esen sekhem esen

"To All the Pyramid Builders
who shall cause the
Construction of this Pyramid
to be good and enduring,
it is they who will be Supreme,
it is they who will be Honored,
it is they who will be Vital,
it is they who will be Powerful,
it is they who will have Control,
it is they who will Rule"

~Pyramid Text of
Nefer Ka Re Pepi II
Utterance 599, Line 1650

Table of Contents

FOUNDATION

1.0. INTRODUCTION

Liberation is defined as separation from adverse forces. In Nature, Physical Liberation is accomplished by separating, or changing and moving location and position from one place to another. In Nature, one of the means that Liberation is fortified is by constructing structures and containers as maintainers of safety, security, and peace of mind for the continuity of the Liberation. Thus, **Liberation is a motivating force in the Creation of Architecture**. Everything that exists in Nature and the Universe has a form, structure, or "Architecture". We define everything in Nature and the Universe to be all Space/Vacuum, Matter/Energy, and Time/Existence. The **Architect** in Nature is the **Energy** of Nature, and the Energy of Nature is the metaphorical **Mind** of Nature. Space is the component part of Nature which provides the container and location in which matter moves and can be used to build. It is the movement of matter by energy through space which creates time. The **Architecture** of Nature includes all of the forms and structures in the Universe composed of **Matter**; from the microscopic to the macroscopic and everything in between. An Architect creating Architecture in Nature occurs when the Energy of Nature is used to form the Matter of Nature to create the perception of division in the Space of Nature. Each perceived division or sub-space of Nature is a reflection of the mind of the Architect that created it. Everything that exists in Nature exists somewhere, in some place, and in some sub-space within space. The structure, and the environment contained within the structure, is a reflection and projection of the mind that created it. Human Architects are one of the many Physical manifestations of the Creative mind of Nature, and the Architecture produced by Architects is their created Space and/or Environment. Once Liberation has been established, **"Establishments"** are created for the perpetuation of the Liberation. Naturally, the Architecture of the Establishment will reflect the mind of the establisher - the Original Architect(s) who initially desired Liberation. Thus, a key indicator of Liberation is the ability to be **Architects of Architecture** which subsequently means **creating and constructing reality**.

The Human development of **Architecture** is intricately connected to the development of **Agriculture**. As Humans transitioned from being Nomadic **"Hunters and Gatherers"** being **controlled by their environment** to becoming **Agriculturalist controlling their environment**, the need to establish settlements and

Architecture naturally arose. If compared to mental processes, the process of going from Nomadic wanderers to Agricultural Settlers can be liked to going from an Inquisitive Curious Wonderer (Student) to a Knowledgeable Explainer (Teacher); i.e. from **Asking Questions** to **Making Statements**. Just as a Species can adapt to an Environment, an Environment can adapt to a species once the species becomes intelligent enough and creative enough to transform the environment.

The Space of Nature is scientifically and mathematically described using a **Three-Dimensional** (**3D**) model with the three dimensions being Length, Width, and Height. In three-dimensional space, after the **Sphere** (**0**) shape, there are **Nine** Polyhedron shapes that can be constructed geometrically. Three of the Nine Polyhedrons, namely **the Tetrahedron**, **the Cube**, and **the Octahedron**, naturally occur in the arrangement of the atoms in **Crystal Structures**. The **Tetrahedron** is considered the base "**Archetype**" crystal structure in that it is the simplest structure that can be constructed in 3-Dimensional space. The Tetrahedron is also known as a **Triangular Pyramid**. As the Archetype of Architectural structures built by Nature in Nature, it is only natural that the Pyramid be examined as a form of Nature Architecture. The Tetrahedron is one kind of Pyramid called a "Triangular Pyramid", however the type of Pyramid that most people are familiar with is the **Pentahedron** or "**Square Pyramid**". Geometrically, there are a multitude of different types of pyramids that can be created by increasing the number of sides of the base including **Pentagonal Pyramids**, on up to **Cones**.

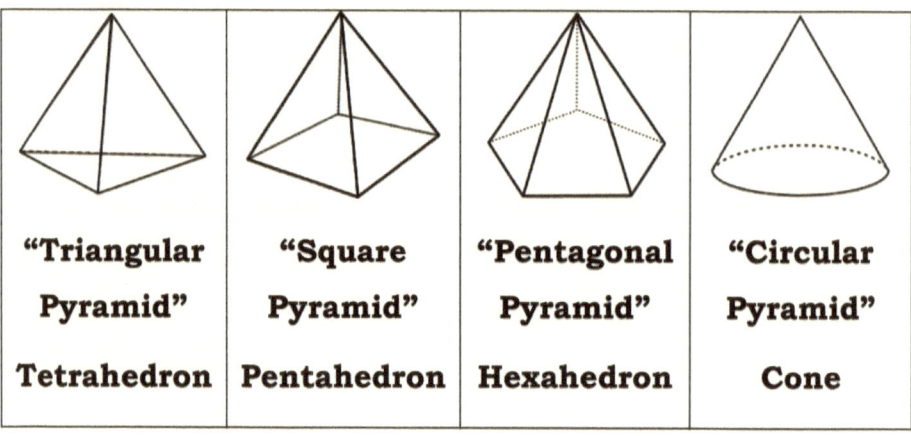

"Triangular Pyramid"	"Square Pyramid"	"Pentagonal Pyramid"	"Circular Pyramid"
Tetrahedron	**Pentahedron**	**Hexahedron**	**Cone**

The Pyramid is perhaps the most iconic and recognizable structure ever constructed by Humans. Pyramid Architecture is a design that has its Earthly origins in Ancient times on the African continent. Thus, Pyramid Architecture, and its many derivatives and functions, is a form of Architecture that reflects and is conducive for being in tune with Nature, and also being in tune with traditional African culture and mentality.

It has been said that **"Man fears time, but Time fears the Pyramids"**. This statement has been used to show the durability of the Ancient Pyramids through time and the ages. As we have previously discussed, we Pyramid Architects know that it is the construction of Architecture by Architects ((movement of matter by energy through space) which creates time, and thus we symbolically recognized that by building Architecture like Pyramids we are simultaneously **Creating Time**.

When you create a new environment you are creating a new reality. All of Existence exists in Space, and each relatively smaller existence exists in sub-spaces of Space and sub-realities of Reality, and each Existence in Space has its own "Time" so to speak. So when it comes to Creating an environment on top of or within an existing environment, you are creating a "New Reality", i.e., a reality that is similar, but not exactly like the reality that it was built upon. Your house, your car, a City, and your body are all examples of this principle. When a "New Reality" is created via Architecture, there are rules and laws that pertain to that created reality that do not necessarily pertain to the reality in which it was created inside of or on top of. The "New Reality" created via Architecture enables the Mind to be taken to "another world". In Quantum Physics and Special Relativity the laws of classical Physics that seem to perfectly describe our world or relative reality, do not work and do not describe the Laws of the **Sub-Atomic** or **Quantum** World. Also, the laws that seem to describe things on the Astronomical level in Space, i.e., Planets, Solar Systems, Galaxies, and Universes, do not necessarily apply to our level or the sub-atomic level. This indicates that there is **Relativity in Reality** and this Relativity if facilitated via Architecture. These different levels are all different Realities, all which are built on the Ultimate Reality which is the world of the smallest Particles of Matter called **"Fundamental Particles"** in Physics (namely Quarks, Leptons (electrons), and Bosons (photons, gluons)).

Architecture not only creates a "New Reality", but it also provides an Environment in which to "Program" and create New Minds. New minds resulting from being created and programmed in and by some form of Architecture can be analogous with **"Artificial Intelligence"**; i.e. a mind created by another mind. Thus, when an Architect creates Architecture that in turns influences and inspires the mind of the original Architect, this is the Creative feedback which has been various metaphorically described as being **"Self-Created"**.

Once Architecture as been built and an Environment has been established within the Architecture, Law is created to maintain the Order established by the mind of the Architect. Laws are also created to regulate the construction of any other Architecture which could potentially usurp the order of the environment. Generally, **"Building Permits"** are usually used to ensure the safety and security of an Architectural structure; however, if the building, or the environment established within the building, can potentially upset the balance of the larger environment in which it will be constructed, then naturally the governing ordinances will not permit the building. Let us consider for a moment the dualistic meaning and implication of the term "Building Permit". Exoterically, a "Building Permit" is permission from a governing body to build a physical structure, but esoterically, "Building Permits" can be used by governing bodies to block the metaphorical "Building" of certain mentalities.

Just as a mind can create an Environment or "Reality", the Environment can likewise effect and influence the mind. The influence of an environment on a mind is the core of the entire **"Nature vs Nurture"** debate. One's "Nature" is the default programming of the Mind, and One's "Nurture" is the programming that can alter your default programming by way of the Environment. The fact that the Environment can effect the Mind (a Reality can effect Intelligence) is illustrated in the laws of Physics called Thermodynamics. The energy of the Environment effects the Energy of the things in the Environment; for example, if you put a Cold Cup of water in a Hot Oven, it will eventually become Hot, and if you put a Hot Cup of Water in a Cold Refrigerator, it will eventually become cold. It is Important to find or Create a Reality that Reflects your Intelligence. It is extremely difficult to conceive of something that is completely unlike anything in your Reality. Thus, the earliest "Artificial Intelligence" was created by Programming Minds by Creating Environments/Realities. So you have to decide, are you an Architect or are you a Resident. Do you create the Architecture, or does the Architecture create you. Do you create your environment and control your Reality, or does your environment effect and control you. The African Creation Energy **"P.T.A."** movement is dedicated to training "Architects", those individuals who are able to use their mind design Architecture so that we as African people can create, build, and construct Structures, Settlements, Establishments, Monuments, Edifices, Palaces, Places, Environments, and Realities that are conducive to our mind. Unfortunately, at the time of the writing of this book, it is observed that many of the

"Nubian" and "Hispanic" descendants of the Ancient Pyramid builders have had to resort to becoming "Construction Prostitutes" and "Handyman-Whores", standing outside of Hardware stores in hopes of someone driving up to them and offering them some money for their building and construction abilities. It is the goal of this book to re-empower the descendants of the Ancient Pyramid Builders to employ their own skills for use on their own projects for the good of their own people again.

Human Architecture and Human Creations are just as natural as any other creation in Nature. If a Beaver's dam made out of wood is a Natural creation and if an Ant's mound made from dirt is a Natural creation, then surely a Human's house made of wood and dirt (bricks) is an equally natural creation. So, in fact there are no unnatural creations. The **Akan** people of **West Africa** expressed the union of Nature's creations and Human creations in the **Adinkra symbol** called "Abode Santaan" which symbolizes the totality of the Universe and the union of Natural and Social creation. The West African Akan symbol of **Abode Santaan** is related to the Ancient Egyptian symbol of the **"Eye of Re"**. There are pyramids all over the world. The Pyramids are awe inspiring and have fascinated and captivated Human minds for generations. There has been a plethora of literature, audio recordings, and videos produced pertaining to the topic of the pyramids. Pyramid subject matter usually ranges from, "Who Built the Pyramids", "Why where the Pyramids built", and "How were the Pyramids built". Whereas the Pyramids of Ancient Egypt have been the source of great fascination over the years, fanatical speculation and fantastic theories about the Pyramids have developed into a somewhat pseudoscientific genre labeled **Pyramidology**. However, rather than speculating on these topics, what distinguishes this book entitled "**ARCH I TECT: How to Build A Pyramid**" is that it actually provides the reader with a method, a means, and a reason to construct their own Pyramid.

We do not propose that the methods presented in this book are the same methods that were used in Ancient times, but rather we

propose a simple, inexpensive, and effective way to build pyramids and pyramid-like structures using innovative and creative techniques. Thus, this book entitled "**ARCH I TECT: How to Build A Pyramid**" is actually a pioneering work that has created (or re-created) a new field called "**Pyramid Technicians**" or "**Pyramidicians**", and **African Creation Energy** will be the **Harbinger** at the forefront of training the **Pyramid Technicians and Architects**, abbreviated **P.T.A.**, of the world as a herald to the resurrection of the craft. Indeed, as the mind of the Architect develops, we shall go from designing and building Pyramids and structures on Planet Earth, to designing and building Planets and Solar Systems throughout the Universe.

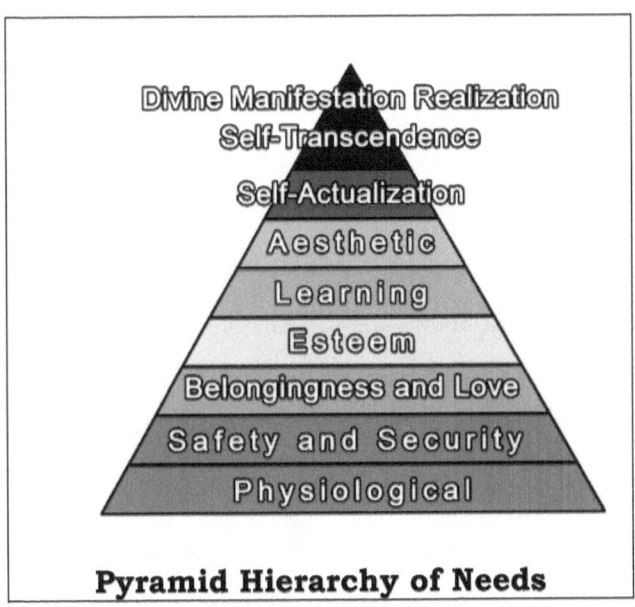

Divine Manifestation Realization
Self-Transcendence
Self-Actualization
Aesthetic
Learning
Esteem
Belongingness and Love
Safety and Security
Physiological

Pyramid Hierarchy of Needs

When we consider the Hierarchy of basic Human needs, after food, water, and the other fundamental physiological needs required to maintain life, the need for safety and security comes into play. One of the ways the basic Human need of safety and security is ensured is through the construction of Architecture. Not surprisingly, the Hierarchy of Human needs is usually depicted in the shape of a Pyramid. The Hierarchy of Human needs is usually listed from the bottom to the top in the order of:

1) Physiological Needs,

2) Safety and Security Needs,

3) Belongingness and Love needs,

4) Esteem Needs,

5) Learning Needs,

6) Creative and Aesthetic Needs,

7) Self-Actualization Needs,

8) Self-Transcendence Needs, and

9) Divine Manifestation Realization.

There has been a multitude of organizations, groups, and literature dedicated to addressing almost every topic listed in the Hierarchy of Human needs except for the topic of the purpose, use, and design of Architecture to realize and actualize the second most important need of Safety and Security. Moreover, proper Architecture creates the environment that is conducive for pursuing and obtaining the other levels of needs. Thus, if the other basic needs are to actually be obtained, it is imperative that our Pyramid of need be constructed in order, hence, this book and the building and construction of the Architecture described herein is literally a gateway to realizing one of the ultimate Human needs of Divine Manifestation. The hierarchy of Human needs is also considered the driving force in **Motivation Theory** in that it is the pursuit of each of the needs which drives Humans to **take action**. Considering that obtaining Safety and Security through the design and construction of Architecture is one of the **Prime Motivators**, this book also serves the purpose of being a Prime Motivator to accomplish **Great Work**. Considering the importance of obtaining **"Food, Clothing, and Shelter"** to the Liberation of a People, this book entitled **"ARCH I TECT: How to Build A Pyramid"** is unique in that it addresses the often neglected topic of **"SHELTER"** from an African perspective, and thus is one of the final pieces of the puzzle needed for the realization of true African Liberation.

The particular purpose that the Pyramids constructed as part of this project entitled **"ARCH I TECT: How to Build A Pyramid"** include:

- To gain first hand experience with various Pyramid building techniques that can be scaled up in order to construct much larger Pyramid Temples, Structures, and Edifices

- To exercise Creative ability and pay homage to the Creator whose likeness we were Created in

- To provide motivation, inspiration, and a living example of what can be accomplished by using AFRICAN CREATION ENERGY

The method for Pyramid Building described in this book is not presented as an explanation into how Ancient Megalithic Pyramids were built, but rather as a reasonable and practical model and method intended to motivate, activate, and inspire the reader to partake in the Art and Craft of Architecture and Pyramid Building. Readers of this book who agree to engage in the science and art of Pyramid building as described herein are taking part in the **"Covenant of the Arch"**. Considering the plethora of practical knowledge that is passed on from one generation to another, and considering that the last Megalithic Pyramid was built just 500 years ago, it stands to reason that there are indeed individuals on the Planet who know precisely how the Ancient Megalithic Pyramids were built. If knowledge about "how to bake a cake" and "how to sew a button" has been able to be passed from generation-to-generation without interruption, then surely knowledge about "How to build a Pyramid" has also been passed on. This work comes almost a year after Archeologist discovered the tombs of Pyramid Builders in Egypt and verified that the Ancient Pyramid Builder were indeed skilled workers and not slaves, and thus this book entitled **"ARCH I TECT: How to Build A Pyramid"** is offered to all of the potential and future Master Builders of the world as a handbook to resurrect the Ancient Architectural Art of Pyramid Building.

The inspiration for the writing of this book entitled **"ARCH I TECT: How to Build a Pyramid"** came after writing a Triad of books which were part of a project collectively called the "African Liberation Science, Math, and Technology Project" (abbreviated the African Liberation S.M.A.T. Project). The purpose of the African Liberation S.M.A.T. Project is to motivate the Creative Energy of African people to Produce, Develop, and create any and everything needed for survival and well being by African people and for African people. Knowing that my ancestry comes from the Ancient Pyramid Builders of Kush, Napata, and Meroe, and knowing that in Ancient times, structures were built to show the power of a particular deity, I choose to embark upon this project of building a Pyramid, and also training and resurrecting the craft of Pyramid Building to show the power of **African Creation Energy** at work. This book and other books written and published by **Osiadan Borebore Oboadee** and the coined term **"African Creation Energy"** are introductory educational tools for a long term goal and mission of growing, cultivating, nurturing, encouraging, supporting, advancing, and promoting African Creativity, Inventiveness, and Ingenuity for the purpose of developing, engineering, forming, formulating, innovating, inventing, designing, building, and creating any materials, structures, machines, devices, systems, and processes needed for survival and well-being by African people for African people. With that said, the importance of building and constructing Architecture is illustrated, described, taught, and emphasized in this book.

2.0. THE ARCHITECTS

The role and responsibility of the Architect is to design a plan that leads to the creation of something. Although the title of "Architect" is primarily used to describe person responsible for the design and construction of a building, the term "Architect" has also been extended to identify individuals responsible for planning, designing, and creating a variety of systems. The word **"ARCHITECT"** was constructed by combining the Greek words **Arkhi** meaning "chief, ruler, or master" and the Greek word **Tekton** meaning "artisan, carpenter, and builder". Thus, an etymological sense of the word **"ARCHITECT"** is **"MASTER BUILDER"**. The word "Architect" is also related to the word "Archon" meaning "Ruler" and also related to the words An<u>archy</u>, Mon<u>archy</u>, and Hier<u>archy</u>. The Architect is one of Humanity's oldest professions. The Ancient role of Architect was also closely related to the modern professions of **Civil Engineering** and **Surveying** in that it was the responsibility of the Architect to design and construct the Human-made structure to be in Harmony with the natural environment. The history of Architects can be traced by studying the oldest Human structures which date back as far as

11000 B.C. Gobekli Tepe, the Megalithic Temples of Malta, the Mound at Newgrange, and the various Pyramids throughout the world are example of some of the oldest Architectural structures that are still standing. These Architectural structures are reflections of and projection from the mind of the Ancient Architects and Master Builders who designed and constructed them.

The Biblical tradition suggests that these great builders were sons and daughters of "God" called **Ghibbore** (Strong's Concordance number H1368) meaning "**mighty men of renown**". The Biblical tradition even suggests that God is the Master **Builder** of the Universe, and Jesus Christ was a Builder or **"Tekton"** (Strong's Concordance number G5045) in Mark 6:3.

The Pyramids were said to have been built from the "Top-Down" in that the "Top" indicates the mind or "mental blueprint" and "Down" represents the physical construction. All master builders and Architects comprehend the metaphorical principle of "**building from the top down**". Master Builders and Architects also comprehend the principle that "**To Build is to Destroy**" meaning that the building or formation of a new creation is the destruction of another creation.

We acknowledge all of the Architects and Master Builders in Nature:

- The Smallest Particles in Nature, "the Subatomic Particles", known as **Bosons** who build using **Leptons** and **Quarks** – Master Builders of everything in our entire Universe

- The Suns and Stars – Master Builders of Solar Systems and Galaxies

- Bacteria and Viruses – Master Builders of Life

- The Coral "Reef Builders" of the Ocean

- Master Builders of the Insect world, The Bee Hive Builders, The tiny Ant Mound Builders, the Termite Mound Builders, and the Spider Web Builders

- Beavers who build dams, and Moles that build Molehills – Master Builders of the Animal world

- The Human Being - Master Builders of Ancient Megaliths, Pyramids, Temples, Edifices, and many other Structures

Some people may feel as if all of the beings listed above are not true Architects, and may feel as if we are *"**Making a Mountain out of a Molehill**"* by giving all of these beings too much credit. However, we acknowledge all of these beings as manifestations of the mind of the

Architect in Nature. Birds use Nature's resources to create nests, Spiders use Nature's resources to create Webs, Bees use Nature's resources to create Hives, and Beavers use Nature's resources to create Dams; all of these are examples of technology to their respective species. The "Human Organism" is in fact a form of technology resulting from smaller organs and organisms working together and applying their respective knowledge to form "Architecture" that would ensure and prolong their survival and well-being. The Architectural Design that began with the creation of the universe, planets, stars, and mountains continues with the Architectural Designs of Nature's creatures in the form of Ant mounds, and Human Skyscrapers, etc.

It is interesting to note that **"Hymenoptera"** (the name for the Insect **Order** of Bee and Ant **Master Builders**) is phonetically similar to the Ancient Egyptian words **"Hemut Ptah"** (an Ancient name for the Order of Pyramid Master Builders). The word "Hymenoptera" means "membrane wing" and comes from the Greek word *hymenaios* meaning "wedlock" or the "unification of the male and the female" (the science of creation). Also, honeybees are a subset of bees in the genus **Apis**, and Apis is the Ancient Egyptian name for the Bull that represented the craftsmen deity **Ptah**.

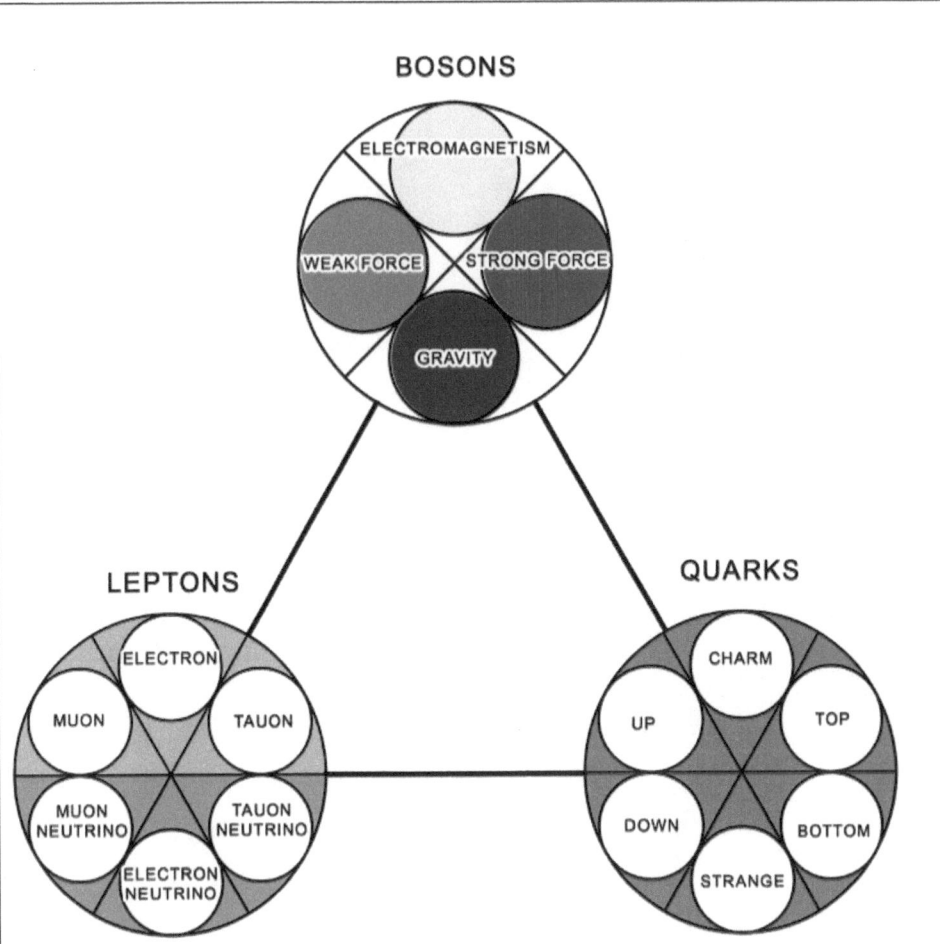

1. The Fundamental Forces - The Grand Architects of The Universe (G.A.O.T.U.). Bosons (Electromagnetism, the Strong Force, the Weak Force, and Gravity) - the Fundamental Forces (Minds) in Nature responsible for Designing, Building, and Creating the universe using the smallest particles of Matter called Leptons and Quarks as explained by The Grand Unified Theory of Everything (G.U.T.O.E.)

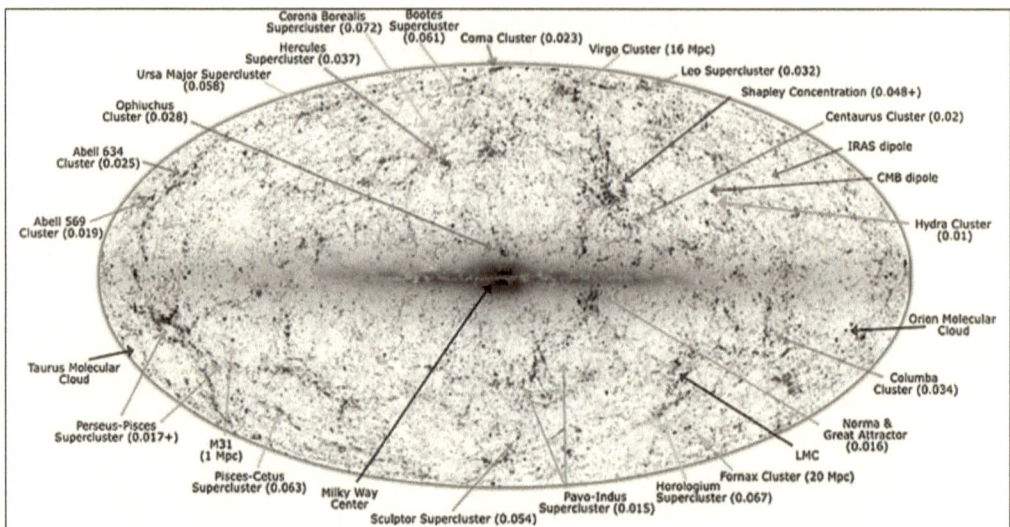

2. The Suns and Stars in The Large Scale Structure of the Observable Universe – Architects, Designers, Builders, and Creators of Solar Systems and Galaxies using Planets, Moons, Asteroids, and other Celestial Bodies and Astronomical Objects in the Universe

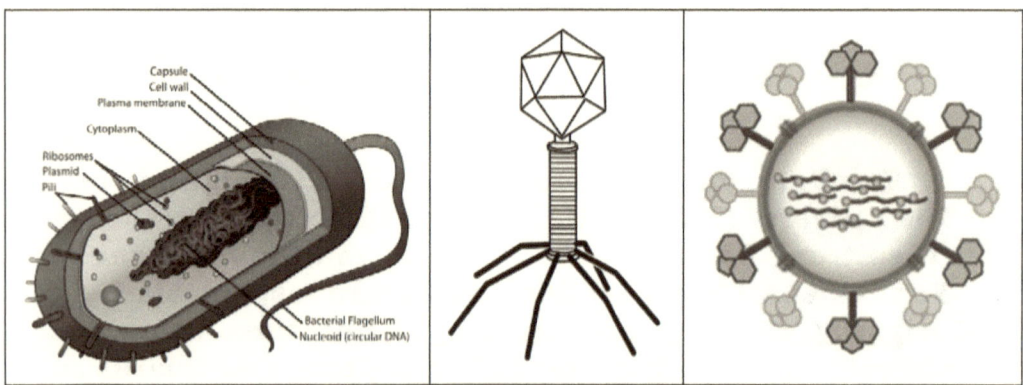

3. Bacteria and Viruses – Architects, Designers, Builders, and Creators of Life

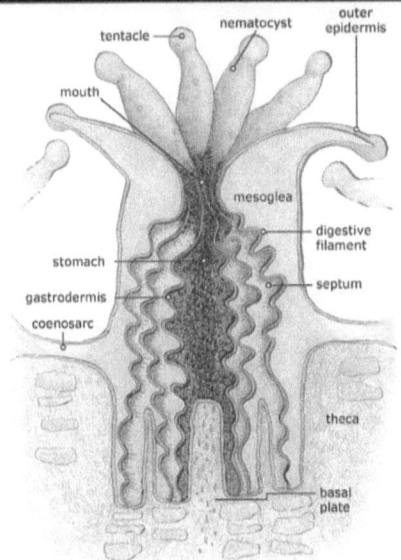

4. Coral – Architects and Builders of Reefs in the Plant and Ocean world

5. Ants and Bees – Architects and Mound Builders in the insect world

6. Beavers and Moles – Architects and Mound Builders in the Animal world

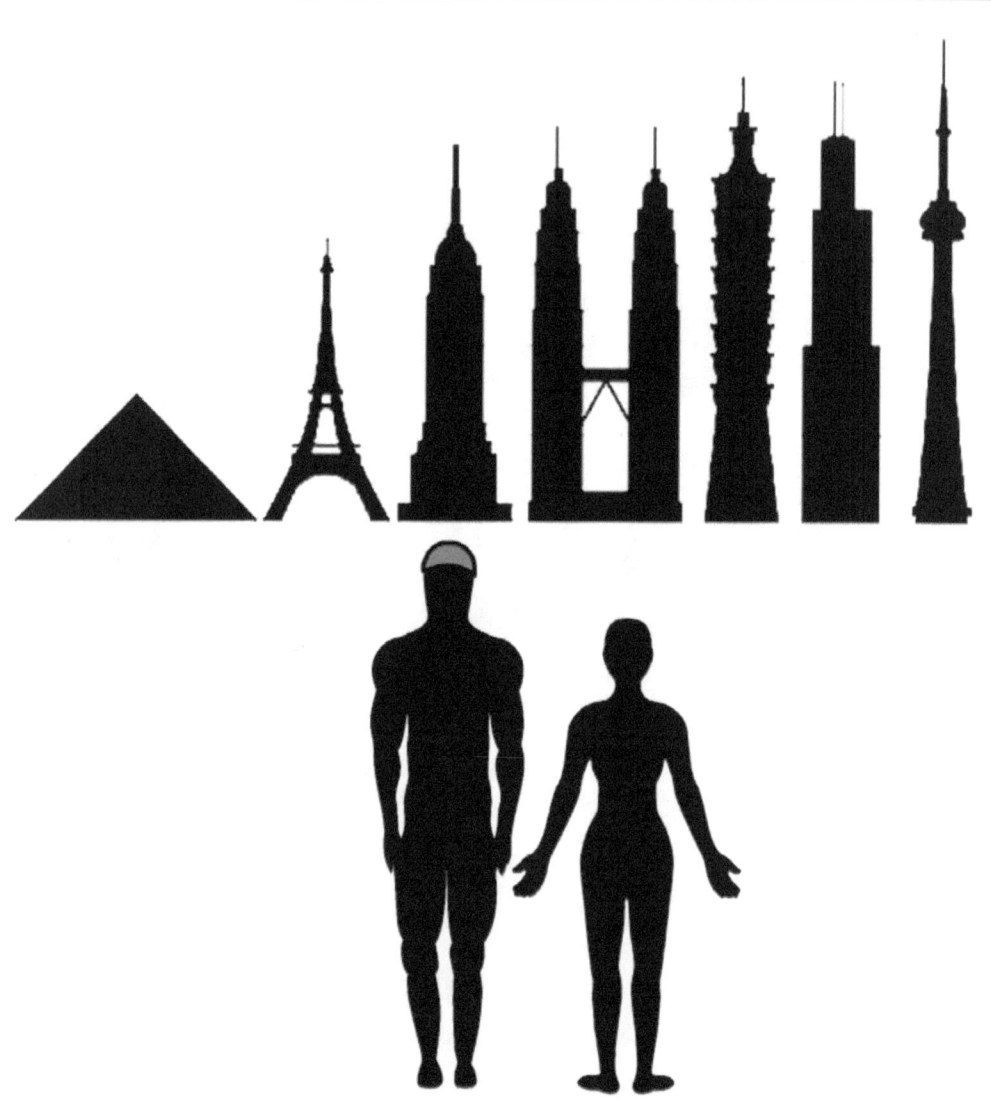

7. Human Beings (Homo Sapiens) – Architects and Master Builders of Mounds, Pyramids, and many other Structures

The Ancient Architects of Egypt usually served as Vizier to the Ruler. It is generally accepted that Narmer or Menes was the first Ruler of the first Dynasty of Egypt. On the Narmer palette which commemorates Narmer's successful unification of Upper and Lower Egypt, he is shown accompanied by his Vizier named Tjaty, the first Vizier of Dynastic Egypt. We also know by name a Vizier named Menka during the reign of the first Dynasty Ruler named Den. Viziers carried the titles "Keeper of the Royal Seal" and **"Master of Works"**.   According to a book entitled *"History of Ancient Egypt"* by George Rawlinson, the Ancient Egyptian name for the "Royal Architect" was **MURKET**. The title MURKET for Royal Architects is also similar to the name for an Ancient Egyptian tool used for Aligning stars and tracking time called a **MERKHET** which meant **"instrument for knowing"**.   Indeed, the MURKET would use the MERKHET instrument to establish alignments of buildings with the stars. At the apex of Pyramid Building in Egypt, the Royal Architects and Viziers were related to the Ruler as a brother, son, or nephew. For example, Ruler Sneferu's oldest son Kanefer served as his Architect and Vizier, and Ruler Khufu's Architect and Vizier was his nephew Hemiunu. However, the inventor of the Pyramid, and the first Architect and Engineer that we know by name was not of the Royal lineage, but was a commoner by the name of **IMHOTEP**.

IMHOTEP

Who was IMHOTEP?

There were many buildings constructed prior to the stepped-Pyramid of Djoser, thus we can conclude that there were many Architects who have lived since time immemorial. However, with the construction of Djoser's Stepped-Pyramid also comes the identity of the earliest Architect that we know by name: **IMHOTEP**. IMHOTEP was the chief vizier to Pharaoh Djoser and the Architect of the Stepped-Pyramid. IMHOTEP was an Architect, Engineer, Physician, Astronomer, Mathematician, Philosopher, magician, mathematician, scribe, sage, astronomer, vizier, Lector-priest, Renaissance-man, and Poet. Due to his many talents, IMHOTEP is considered one of the first multi-genius **Polymaths** (wise sage) in the world. IMHOTEP was born on May 31 and estimated to have lived for 55 years from 2655 BC to 2600 BC (although estimates vary). IMHOTEP's abilities as an Architect in designing the Stepped-Pyramid, and his talents as a Physician and Healer led to his status being elevated to that of a Demigod and Demiurge amongst the Egyptians. The name "I-M-HOTEP" means **"coming into peace"** or **"coming into satisfaction"**, and thus to be in a state of "I-M-HOTEP" or "peaceful satisfaction" with one's work would be the goal of any creator, craftsmen, artisan, or Architect. IMHOTEP was a commoner by birth, but rose through the ranks as a **self-made nobleman** due to his remarkable talents. IMHOTEP held many titles and was cross-trained in a variety of disciples represented by various Ancient Egyptian deities:

- As a **"High Priest of the Sun God RE"** called the **"Greatest of Seers"**, IMHOTEP was an **Astronomer**.

- As a **"Son of Ptah"** or **"student of the craftsmen"**, IMHOTEP was a **builder**, **carpenter**, and **Architect**.

- As **"one of the Ibis and the Pen"**, IMHOTEP was student of **Tehuti** and a **Mathematician**.

- As a **"Grandson of Khnum"**, IMHOTEP was a **sculptor**, **molder**, and **Creator**.

- As a **"son of Sekhmet"**, IMHOTEP was a **Physician**.

It is believed that Imhotep's family was from the Southern Egyptian city named Inerty (Naga Gebelein) or possibly Ankhtown, a suburb of Memphis. In the book *"Black Genesis"* by Robert Buval and Thomas Brophy, it states that Imhotep was a descendant of Sub-Saharan Africans, and paid homage to his African Ancestral Architectural tradition by designing buildings with similar astral alignments as the older megaliths found at Nabta Playa (*"Black Genesis"* page 251).

The works of IMHOTEP were so great, and the stories about IMHOTEP so fantastic, that there has been debate as to whether or not IMHOTEP was indeed a real person. One of the primary reasons why the empirical existence of IMHOTEP is disputed comes from the fact that neither a mummy nor a tomb of IMHOTEP has ever been found. Some people claim that "IMHOTEP" was not only the name of an individual Architect, but also the name used by a "brotherhood" or "Company" of Master Builders. It is interesting to note that in Ancient Egypt, the word for **"builder"** or **"mason"** was **KWDU** which is phonetically similar to the **West African Akan** name **KWADWO** which means **"Spirit of Peace"** and one of the names of the Ancient Egyptian brotherhood of builders or **KWDU** was called **HOTEP** (meaning **peace**). And so it is said in the Christian Bible: "**Blessed are the peacemakers, for they shall inherit the Earth**" (Matthew 5:9).

During the reign of King Djoser, there was a drought in Egypt that lasted for 7 years called the **"7 Year Famine."** Imhotep became the Vizier to Djoser due to the solutions Imhotep provided to solve the "7 Year Famine." The artifact recording this event is called the **"The Famine Stela"**. On the Famine Stela it states:

*"In order to find a solution to the 7 year famine that was plaguing Egypt, I [Djoser] consulted **one of the staff of the Ibis, The chief lector-priest** of **Imhotep, Son of Ptah**. [I Djoser asked Imhotep] Where is the origin of the Nile River and which Deity controls it? [Imhotep responded:] "Let me refer to the Scrolls in my Library and I will return with an Answer. When Imhotep returned, he told Djoser that the Nile Originated in the Southern Part of Egypt in a city named **Yebu (Elephantine)** and the Deity of the temple in Yebu is named **Khnum**.*

After learning that **Khnum** was the deity at the source of the Nile who could end Egypt's 7 Year Famine, Djoser says that Khnum appeared to him in a dream telling him to visit the Temple of Khnum in Yebu so that Khnum could teach Djoser and Imhotep about the materials needed to build a Pyramid. The Famine Stela states Khnum appeared to Djoser in a dream saying:

*"I am **Khnum**, your **Creator, Maker**, and **Fashioner**! I come giving you knowledge of many stones that were not previously known, but can be used for Building Pyramids, Rebuilding Temples, and making Statues. For I am the master who Creates, **I am self-created**, the first to **rise out of the chaos of Nun**, as **Ptah-Nun**, father of the Neteru."*

Upon visiting the **"Temple of Khnum"**, Imhotep received revelation and inspiration on a method to fabricate Limestone blocks as it is recorded on The Famine Stela:

> *"There is a massive mountain in the South-Eastern lands near the Nubians. This mountain is the source of various precious stones, ores, and minerals that can be quarried. All the materials needed to build Pyramids and Temples can be found there. Imhotep give Khnum an offering and then Khnum reveals to Imhotep a list of precious stones, minerals, and other elements and materials that can be used for building Pyramids."*

This list of Pyramid Building materials that was given to Imhotep by Khnum has been called **"The Revelations of Imhotep"**, and is considered "sacred" amongst craftsmen. The use of these materials to cast and mold the blocks used to build the Pyramids is the way by which the blocks of the Pyramids were carved "without the sound of a hammer, nor the sweat of a brow". Although the list of materials was given to Imhotep by Khnum, the deity of the Craftsmen in Ancient Egypt was named **Ptah**.

Both the principle of Ptah and the principle of Khnum are just two of many aspects of **African Creation Energy** expressed via allegorical stories. Since Ptah had called creation into being, he was considered the god of craftsmen, and in particular stone-based crafts. Due to the use of stones to build tombs, Ptah also became associated with reincarnation. Another deity named **Seker** or **Sokar** was also god of craftsmen who specialized in metal working (**blacksmiths**). In fact, the area of Egypt known as **Saqqara** where the first stepped-pyramid was built is named after the deity Sokar. It is also interesting to note that the Mesopotamian Pyramid

Above: African Creation Deity PTAH, Father of IMHOTEP

called the **Ziggurat** is phonetically similar to **Sokar** and **Saqqara**. Since Ptah and Sokar were Gods of the Craftsmen, both deities were merged into one deity named **Ptah-Sokar** or **"Path Seeker"** who was considered the personification of the **"Midnight Sun"**. Eventually, Ptah-Sokar was merged with another reincarnation deity named **Asar** becoming the Master Builder "Asar-Ptah-Sokar" or **"A. Path Seeker"** who was depicted as a dwarf.

Above: The Hawk-Headed Blacksmith Deity Sokar depicted riding the "Feathered Serpent" during the 4th and 5th Hour of the "Underworld" from the Tomb of Thutmosis III.
The Craft Sokar rides is called the HENNU

Once elevated to the level of divine, to the craftsmen Imhotep represented a proto "Christ-like" figure who was the **"Son of Man"** and the **"Son of God"**. Imhotep's "God Father" was said to be **Ptah**, and Imhotep's "Human Father" was named **Kanofer** who was believed to have been an Architect himself and worked for King Djoser's father Khasekhemwy. At this point, it is necessary to discuss the duality of the Architect being called a "son of man" and a "son of God". For every created thing, every constructed structure, there is an Architect (the mental creator) and a Builder (the physical creator). Sometimes the Architect and the Builder are one in the same, and sometimes they are not. Metaphorically speaking, the "mental creator" is considered "God" and the "physical creator" is considered "Man". We Human beings are all physical products of our mother (ovum) and father (semen). However, our mind and mentally is created over time by our environment, life experiences, and many teachers. If there is a philosophy, ideology, school of thought, or teacher that is primarily responsible for our mind and mentality, then metaphorically speaking, that philosophy, ideology, school of thought, or teacher would be the creator of your mentality or "God". So, while our physical body is created by **Semen**, our mind and mentality is created by **Seminars**. This is what is meant by Architect's duality of being "Son of Man" and "Son of God", and in the case of Imhotep, his physical father was Kanofer and his mental father was Ptah. While Ptah has "fathered" many Architects throughout history, the physical lineage of Imhotep can be traced from father to son from 3000 BC to about 400 BC. Some of the Architects in this lineage are Ka-Nofer, Imhotep, Ra-Hotep, Nofer-Mennu, Pepi, Nass-Hunu, A'ahmes, and Khnum-Ib-Re.

After the deification of Imhotep, it was said that his mother Khredu-Ankh (meaning child of life) was a daughter of a **Ram** deity from the city of **Djedet** (also known as **Mendes**) named **Ba-Neb-Djed-et** (meaning "Soul of the Master of Stability"). This Ram deity from the city of Mendes is where the more recent occult symbol of "**Baphomet**" or "**the GOAT of Mendes**" is derived. However, "the GOAT of Mendes", Ba-Neb-Djed-et is actually a symbol of **Creativity**, and is another form of the Ram-headed Creation deity named **Khnum**. The supreme Creator deity of the Bambuta Pygmies in Central Africa is the deity named **Khonvoum**. In the Bambuta Pygmy's creation story, it states that Khonvoum created the different races of people from **different types of clay**. The African creation deity **Khonvoum** was called **Khnum** in ancient Egypt, and it was said that Khnum created, shaped, molded, and fashioned the bodies of humans out of **clay** on his **Potter's Wheel**. The name Khnum means "**to build**, unite, or join" and he was also considered the "divine potter or molder", "master of created things", and the father of "**Magic**" (**Heka**). It is said that it was Khnum who gave Imhotep and the Ancient Egyptian Pyramid builders the knowledge to mold giant Limestone blocks, thus eliminating the need to quarry and drag large stones.

Khnum or Khonvoum

African Creation deity called the creator, potter, builder, and molder

At the time when Imhotep was called by Djoser to solve the drought problem in Ancient Egypt, Imhotep had already obtained the titles of "Son of Ptah" and "Lector-Priest of the staff of the Ibis". The title "Son of Ptah" means that Imhotep was already a trained and skilled craftsman. The title "**Lector-Priest of the staff of the Ibis**" means that Imhotep was working in the role of a Teacher and Lecturer called a "**Kheri-Heb**" as a student of the Ibis-headed deity named **Tehuti**. Imhotep's name was often grouped with such powerful deities as Tehuti, the God of Wisdom. Tehuti (also called **Djehuti**, **Zehuti**, or Thoth) was the Egyptian god of Wisdom, Writing, Magic, Sciences, Math, and measuring time. Tehuti was a lunar deity and often depicted as a man with the head of an Ibis bird or as a Baboon. In the Egyptian mythology, Tehuti was the Husband of **Ma'at**, the Egyptian Goddess symbolic of Truth, Justice, Order, and Divine Wisdom. Tehuti's other female counterpart was the Egyptian Goddess **Seshat**. Seshat (whose name meant "she who scribes") was the Egyptian Goddess of wisdom, knowledge, writing, **architecture**, astronomy, astrology, building, mathematics, and **surveying**. Tehuti was responsible for transforming creative thoughts into action.

Tehuti was the African Egyptian deity of Thought, Knowledge, wisdom, writing, sciences, and mathematics.

Tehuti's wife Seshat bore the title "The Enumerator" and **"Lady of the Builders"**. Seshat was depicted dressed in a leopard skin, and to wear a leopard skin in ancient Africa indicated a relationship to the **builders** and **craftsmen**. The Egyptian Goddess principle of Seshat was present when the surveying and Architectural design of Egyptian pyramids and temples was being done to align with certain celestial bodies in the cosmos. In order to ensure a good construction, Seshat was present at every Ancient Egyptian **"ground breaking"**, **"foundation laying"**, and **"corner stone setting"** ceremony for any building project. The first and most import "ritual" in any construction project involving Sheshat was called "pedj shes" meaning **"stretching the cord"**. The "Stretching the cord" ritual used a rope to mark the borders of a future construction site. It was the responsibility of Seshat to reflect the order of the cosmos when designing buildings to be aligned with the stars in order to bring **"God's Kingdom to Earth"** (as above, so below). In addition to the "Stretching the cord" ritual, seven other import "rituals" that occurred during the construction of Pyramids or temples were Digging the foundation trench, Molding the first bricks, Pouring sand into the foundation trench, Burying foundation deposits, Initiating construction, Blessing the completed building, and the Dedication Ceremony once the construction was completed.

Seshat was the African Egyptian Goddess of wisdom, knowledge, writing, architecture, astronomy, astrology, building, mathematics, and surveying.

Imhotep began his studies at a very young age, and was likely just a teenage when he was called by King Djoser to help solve the "7-year-drought" problem and design the first step Pyramid. The Step Pyramid of Djoser built a Sakkara was designed by Imhotep as a way for Djoser to perform the rituals of the jubilee **30-year Djed festival**. Imhotep likely lived up to the time of the ruler named Sneferu either dying just before or during the reign of Sneferu. It is likely that during his lifetime, Imhotep designed, influenced, and worked on many other projects other than the step Pyramid of Djoser. As the Architect of the first step Pyramid, Imhotep was likely instrumental in the training of future Architects and Pyramid builders. It is believed that the plans for Sneferu's Pyramids, the Giza Pyramids, and many other temples built after Imhotep's death were conceived by Imhotep during his life and passed on to the Architects who succeeded him to be built in the future. Consider that after Imhotep's death, Sneferu and his Architects **Kagemni I**, **Nefermaat I**, and **Kanefer** go on to build 3 Pyramids: the Pyramid at Meidum, the Bent Pyramid, and the Red Pyramid. The Architects after Imhotep were usually in the blood-line of the ruling Pharaoh; this was the case for the purported Architect of the Great Pyramid of Giza built for the Pharaoh Khufu was named **Hem-Iunu**.

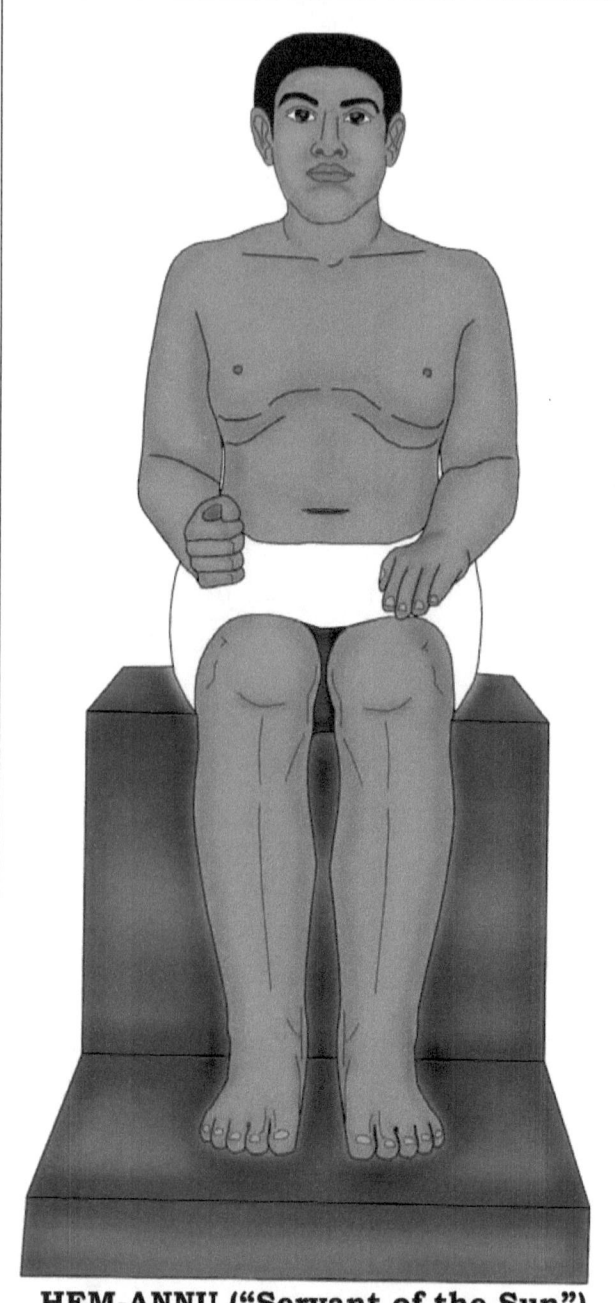

**HEM-ANNU ("Servant of the Sun")
also called Heman, Architect of the
Great Pyramid of Giza called
"Akhet Khufu"**

Hem-Iunu is considered the Architect of the largest Ancient Pyramid on Earth known as the Great Pyramid of Giza. Hem-Iunu, whose name means **"Servant of the Sun (Iunu or Annu)"** is believed to have been born around 2570 BC as was the grandson of the Pharaoh Sneferu and the cousin to the **Pharaoh Khnum-Khufu** for whom the Great Pyramid of Giza was built. For his great Architectural works, a statue of Hem-Iunu was carved and placed in his mastaba tomb near Khufu's Great Pyramid. The statue of Hem-Iunu was discovered in poor condition and the head of the statue was in rubble completely damaged. The statue of Hem-Iunu has been reconstructed and now sits in a museum in Germany. Also discovered in Hem-Iunu's tomb were several papyrus that have collectively been named "The Diaries of Hemiunu" which give great insight into the mind and motivation of the Architect who built the Great Pyramid of Giza called **Akhet-Khufu**.

In some cases the Pyramid Architect and the Pharaoh were one in the same; this was especially true for the 12th Dynasty Nubian Vizier **Amenemhat I** who later became pharaoh. It is also believed that many of the Nubian Pharaohs of the 25th Dynasty were Architects of their own

Pyramids. It is no coincidence that Architectural know-how was prevalent in the area of Nubia, and the philosophy associated with the Architect deity Ptah called **"The Memphite Theology"** was resurrected by the Nubians. Memphis is the name of the first and Capital City in Ancient Egypt. Memphis is also called by the name **Noph** (Strong's H5297, coming from **Na-Ptah** meaning "people of Ptah") in the Judeo-Christian Bible, thus the "Memphite Theology" is also known as the **"Philosophy of Noph"** or "NoopooH". Our modern knowledge of the Memphite Theology comes from an artifact dated to be from 700 BC called **"The Skabaka Stone"** attributed to the Southern Nubian Pharaoh **Shabaka Nefer-Ka-Re** from **Napata** and the Kingdom of **Kush** or **Cush**. Although the physical artifact of the Shabaka Stone dates back to 700 BC, the wording and text contained on the Shabaka stone is much older. The earliest dates for the text of the Shabaka stone place the philosophy to be older than 3200 BC. Even the Judeo-Christian Bible suggests that Africa and Nubian was the origin of Architectural knowledge. In 1 Kings 7:13-14, the Architect and Builder of **Solomon's Temple** is named **Hiram** from the tribe of **Naphtali**, and it is theorized that the origin of the word "Naphtali" is related to Napata, Na-Ptah, and Noph mentioned earlier. In Genesis chapter 10 verse 8, the Judeo-Christian Bible also suggests that great Architectural knowledge came out of Africa when it states that **Nimrod**, the Architect and builder of the **Tower of Babel** was a son of **Cush**, and Cush was the name of the Nubian Empire containing Napata, and also is a Hebrew word meaning "Black" (Strong's Concordance number H3568). The Judeo-Christian Bible also suggests that Nimrod's uncle who was named **Put** was symbolic for the Ancient African **"Land of Punt"**. Although interpretation on this point varies with some Biblical scholars suggesting that Put is associated with the North African Libyans, the fact still remains that the Southern African **Land of Punt** was called **Ta-Netjer** meaning **"God's Land"** by the Egyptians. This point combined with the fact that the Ancient Egyptian word for **"Temple"** was **"Hut-Netjer"** literally meaning **"God's House"** and the Ancient Egyptian word for **Vizier** was **"Hem-Netjer"** literally meaning **"God's Servant"** suggests that the Ancient Egyptians acknowledged the southern Africans as the source of Architectural knowledge. The Philosophy and Seminars of Ptah "gave birth" to Pyramid Builders all over the world. The epigraphic evidence suggests that the **Olmecs** and Meso-American Pyramid Builders share many similarities in Headdress, Culture, Deities, Phonetics, Architecture, Mathematics, and Calendars with the Egyptians and Nubians.

3.0. THE PYRAMIDS

Pyramid shaped structures have severed a variety of purposes throughout history. Some of the functions of Pyramids throughout history have been Spiritual Temples, Tombs, Edifices, Observatories, Laboratories, Residences, Power Generators, and Learning centers just to name a few. The Ancient Egyptian word for "Pyramid" was **"MIR"**, and it is quite possible that the English word "Pyramid" is derived from the Ancient Egyptian Phrase **"PA RE MIR"** which would mean **"The Sun MIR"** or "The Sun Pyramid". Consider the coincidence that one of the functions of the Pyramids or MIR was to be used as an Ancient Astronomical observatory, and there was an actual modern observatory constructed named the **"MIR Space Station"**. Also the word "MIR" can be found in the name of the reflecting glass "Mirror", and the Ancient Egyptians constructed Pyramids or MIR to reflect the celestial sky, "As Above, So Below". The English word "Pyramid" is also phonetically composed of the prefix **"Pyro-"** meaning "Fire" and the suffix **"-Mid"** meaning "middle". The phonetic analysis of the English word Pyramid seems to indicate a central or "middle" point of Energy or "fire" which is associated with the Pyramid.

THE GENESIS AND EVOLUTION OF THE PYRAMID

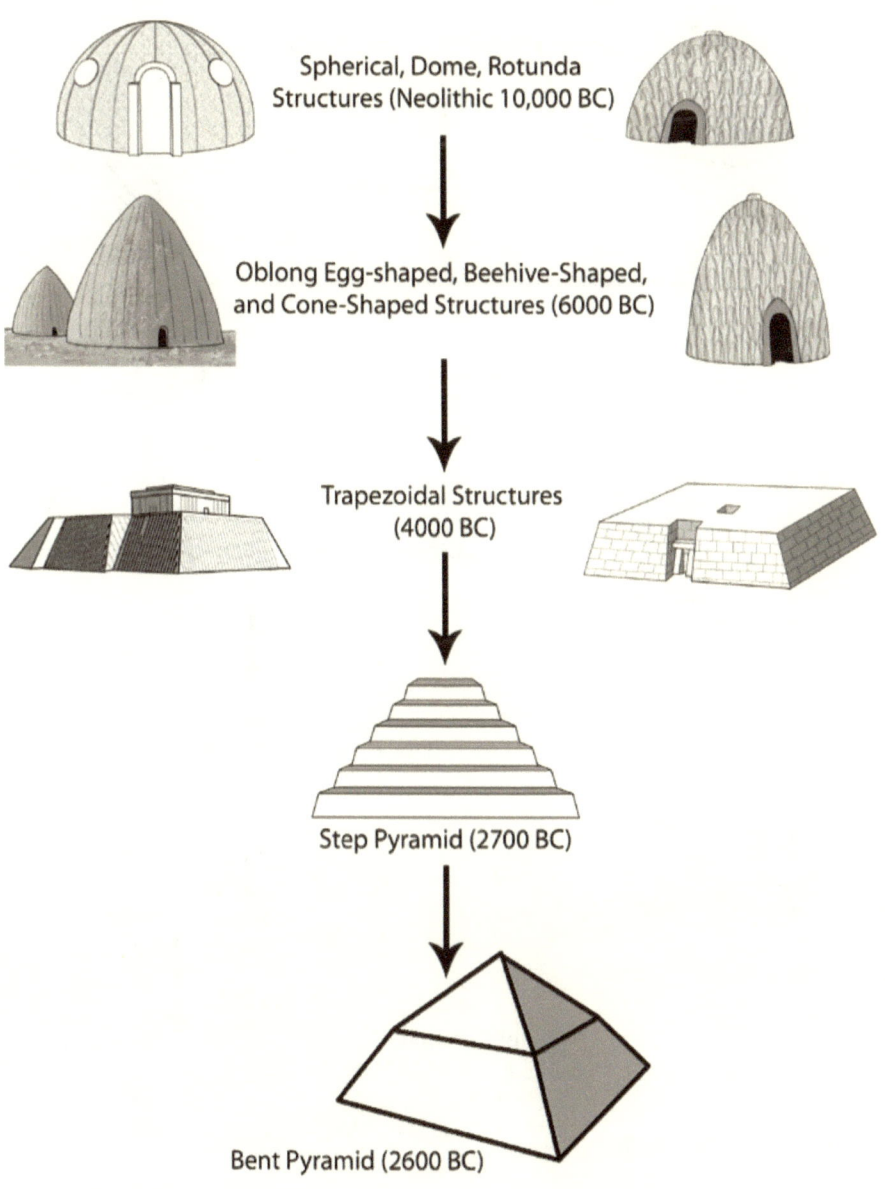

Spherical, Dome, Rotunda
Structures (Neolithic 10,000 BC)

Oblong Egg-shaped, Beehive-Shaped,
and Cone-Shaped Structures (6000 BC)

Trapezoidal Structures
(4000 BC)

Step Pyramid (2700 BC)

Bent Pyramid (2600 BC)

THE GENESIS AND EVOLUTION OF THE PYRAMID
(continued)

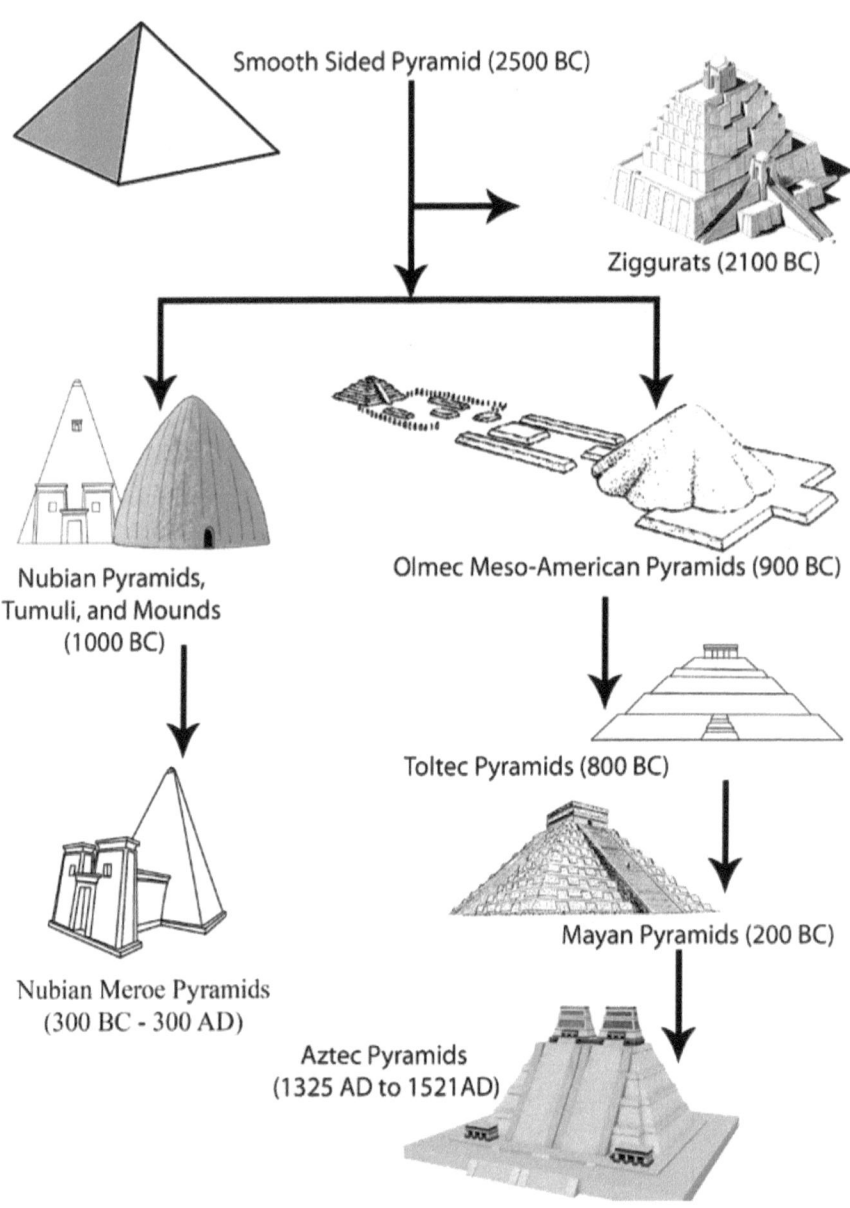

Smooth Sided Pyramid (2500 BC)

Ziggurats (2100 BC)

Nubian Pyramids, Tumuli, and Mounds (1000 BC)

Olmec Meso-American Pyramids (900 BC)

Toltec Pyramids (800 BC)

Nubian Meroe Pyramids (300 BC - 300 AD)

Mayan Pyramids (200 BC)

Aztec Pyramids (1325 AD to 1521AD)

Just as Organisms grow, change, and evolve over time, Mechanisms and Technologies also grow, change, and evolve over time. Architecture and Pyramids both being forms of Technology have also undergone a cycle of change, from a point of conception and birth, to growth and development, evolution, to death and decline, and eventually to resurrection (which this book has initiated). To trace the Genesis of the Pyramids, we must trace the Genesis of Human Architecture in General. As we mentioned earlier, after the Sphere, the Tetrahedron (or Pyramid) is the next Geometric shape constructed by the smallest particles in Nature. Remarkably, the change in the Geometry of human Architectural Structures has undergone the same development. Archeological evidence points to the earliest Humans structures being constructed in the shapes of **Circles** or **Spheres**. In a book entitled "*Temples of the African Gods*", the earliest discovered evidence of Human Architecture is an **Ancient African Metropolis** near **South Africa** that has been dated using Archaeo-astronomy to have been built between **200,000** and **160,000 BC**. The buildings of this 200,000 year old African Metropolis appear to have been **Spherical**, and the city plan was based on a **Circular geometry**. Similar circular arrangements can currently be found in the fractal patterns of various Huts and villages found throughout Africa today. Circular and

Spherical Architecture similar to the 200,000 year old African Metropolis can also be found in an area known as **Göbekli Tepe** which has been dated to 9000 BC.

Amongst the Neolithic African civilizations that pre-dated Dynastic Egypt, we find evidence of Spherical, and Canonical shaped Architecture in the form of living dwellings, **Tumuli** for burial and Circular city plans. In the

Example of Dome shaped Architecture of "God's House" found in "God's Land" called Punt

Papyrus of Hu-Nefer, it states: "*We [the Ancient Egyptians] came from the beginning of the Nile where God Hapi dwells, at the foothills of The Mountains of the Moon.*" The beginning of the Nile is in central Africa in Uganda and Ethiopia. In Ancient times, this area was known as **"The Land of Punt"** or **"Ta Netjer"** meaning "God's Land" or "land of the Ancestors". The houses in "God's Land" were naturally "God's House". A wall carving created after an expedition to the Land of Punt during the reign of Queen Hatshepsut depicts "**Beehive**" and dome shaped Architecture raised on stilts above water. The Egyptian word for "God's House" was "Hut Netjer", which is translated as "Temple", thus, the original Temples were rotunda shaped buildings.

In Nubia, we find evidence of oblong Egg-shape burial Tumuli. In Ancient and modern Times, Nubians built these dome shaped tombs called **Gubbas** or **Qubbas** to represent the **"primordial mound"**. This Dome and Arch Architecture found in Nubia has been called **"Nubian Vaults"** and **"Nubian Domes"**. The Architectural influence of the **Nubian Arch** can be found in Egypt in The **Ramesseum** of Pharaoh Ramesses II. As residential dwellings, the Spheres, domes, and cone shaped architecture applied the science of Thermodynamics to construct Architecture that could remain cool in the hot climate of the desert. In Architecture with high dome or canonical ceilings, the heat rises to the top of the structure making the ground-level area of the building cool and comfortable; the **Musgum** people of Central Africa build dwellings applying this science. The science of constructing Dome and Cone shaped Architecture to create a cooling effect has also been used to create what is called a **"YAKHCHAL"** or "desert refrigerator".

Gubba Tumuli

Musgum House

Yakhchal

As the Pre-dynastic Egyptian culture began to wane and the dynastic Egyptian culture began to wax, a transition was made from Dome and cone shaped architecture **"Built on the Circle"** to Trapezoidal and Rectangular shaped architecture **"Built on the Square"**. Thus we observe the Trapezoidal shaped tombs called **"Mastabas"** used by early Dynastic Egyptian rulers, and also the Trapezoidal and rectangular shaped **"Temple of Anu"** built in Mesopotamian city of Uruk.

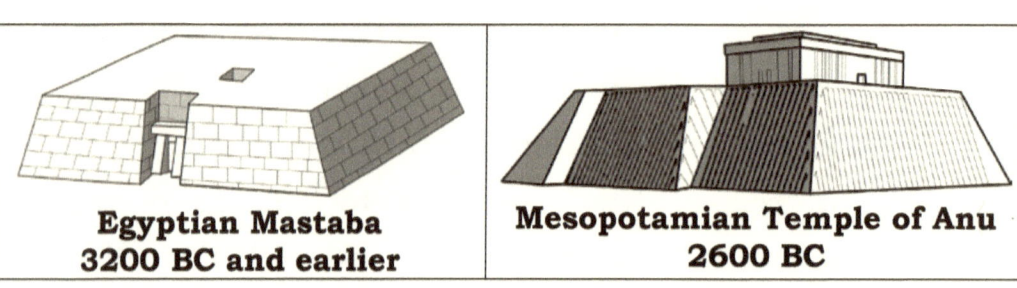

| **Egyptian Mastaba** **3200 BC and earlier** | **Mesopotamian Temple of Anu** **2600 BC** |

Although the "Temple of Anu" is usually called a Ziggurat, we observe that it consisted of only one trapezoidal layer topped with a temple, thus it was not a multi-tiered Ziggurat. The Mastaba tombs of Egypt were also constructed in similar forms as the Mesopotamian "Temple of Anu" topped with a temple. However, the Egyptian Masatabas pre-date the Mesopotamian "Temple of Anu". Thus it is likely that the Mesopotamian builders of the Temple of Anu were inspired by observing Egyptian Mastabas. The "Temple-topped" Mastabas were a pivotal point in the evolution to Pyramid building. In the design and construction of the first Pyramid that we know of, Djoser's Step Pyramid, Imhotep was likely inspired by two unique Mastabas that had been previously constructed. The Mastaba of a **Vizier** named **Nebitka** who served the Egyptian **1ˢᵗ Dynasty** rulers named Den and Anedjib around 3000 BC was built of mud brick, and the sides of Nebitka's Mastaba had eight steps rising at an angle of 49°. Modern Archeologist have named Nebitka's tomb, **Mastaba 3038**. The other Mastaba that likely inspired Imhotep's design of the Step Pyramid came from the Mastaba of Djoser's own father **Khasekhemwy**. Archeological expeditions of the Mastaba of Khasekhemwy indicate that this mastaba was topped with a shrine or a temple. Thus, it is likely that the design of Nebitka's Mastaba and Khasekhemwy's Mastaba and possibly many other structures influenced Imhotep's design of the first know Pyramid. However, the first Pyramid was not always intended to be a Pyramid. Initially, Djoser's tomb at Saqqara was intended to be a Mastaba. After the Mastaba was constructed, an enclosure wall was built around the Mastaba. However, the wall was built so high that the Mastaba was not visible outside of the wall, thus the need for Imhotep to "stack" several Mastabas on top of each other so that the tomb could be visible outside of the wall. The Step Pyramid initially had 4 tiers, but was expanded to 6 tiers.

Djoser's Step Pyramid at Sakkara

The construction of Djoser's Step Pyramid was perhaps the most significant event in the history of Architecture. Naturally, the rulers who succeeded Djoser would want to build similar or better monumental structures. The only way to top Djoser's Step Pyramid would be to build a Pyramid even taller or perhaps, with smooth sides. Much is said of Imhotep's success with Djoser's Step Pyramid, but there is seldom any mention of one of Imhotep's failures. A pyramid known as the **"Buried Pyramid"** which is believed to have been designed by Imhotep for the Pharoah named **Sekhemkhet** was intended to be larger than Djoser's Pyramid, however its design was unstable and the project was abandoned. Later, an **Apprentice of Imhotep** named **Huni** who served during the reign of Djoser would eventually become Pharoah, and make the first attempt at a "Smooth-sided" Pyramid and initiate a blood-line of Pyramid Builders in Egypt. The **Pyramid at Meidum** was started by Huni and finished by his son **Sneferu**. The Pyramid at Meidum was constructed to be taller than Djoser's Pyramid with **9 Layers** at its core and covered with smoothed blocks to make it the first known **"smooth sided pyramid"**. Over time, the smooth-sided blocks have collapsed leaving only the stepped-core which still stands till this day.

Huni – Pyramid at Meidum

Following his father Huni and just after the death of Imhotep, came perhaps the greatest Pyramid Builder named **Sneferu** and his Architects and Viziers **Kagemni**, **Nefermaat I**, and **Kanefer**. Sneferu is considered perhaps the greatest Pyramid Builder because of the three Pyramids that are attributed to him. Sneferu is credited with the completion of the **Pyramid at Meidum**, and the construction of the **"Bent Pyramid"** and the construction of the **"Red Pyramid"** at a time when Pyramid Building was still a new concept. It was under the reign of Sneferu that Pyramids made their greatest evolution from stepped structures to smooth-sided structures.

In the writings left by Sneferu on an artifact known as the "**Palermo stone**", he states that he made expeditions into **Nubia** to acquire a **crew of Pyramid Builders**, and modern Archeology has confirmed that the Pyramid Builders were indeed hired talented craftsmen and not slaves. This combined evidence means that the greatest, most enduring, feats whose construction baffles modern scientist and engineers was design by **skilled African craftsmen**.

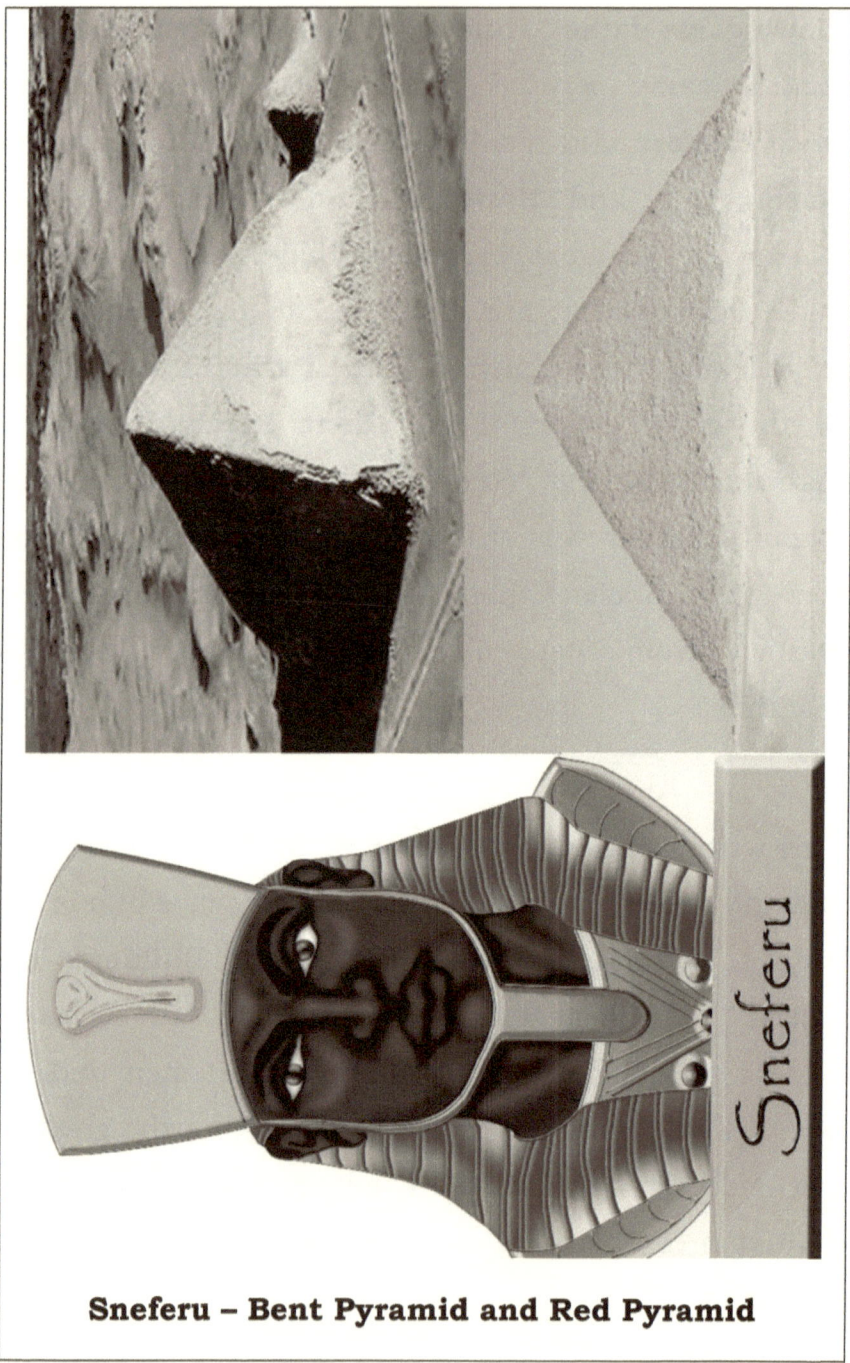

Sneferu – Bent Pyramid and Red Pyramid

In order to comprehend the evolution in Pyramid Building that occurred during the reign of the Pharaoh Sneferu, we must first comprehend the Philosophies that motivated the construction of Pyramids. The philosophy of the actual builders and craftsmen in Egypt was the Philosophy of the deity Ptah centered at Memphis and their sacred symbol was the "Primordial Mound" which was said to represent "Ptah" rising from the primordial waters. The philosophy of the Astronomers was the Philosophy associated with Atum or RE, the son of Ptah, centered at Heliopolis or Annu and their sacred symbol was the "Sun" (star) or light that sat on top of the Primordial Mound symbolized by a Phoenix bird or BenBen stone. Obviously the philosophy of Ptah and the builders was necessary to construct Pyramids, but the Philosophy of Atum-RE and the Astronomers was necessary to align the construction of the Pyramids with the stars. Both the Craftsmen and the Astronomers had angles which they considered "sacred". The Craftsmen favored angles 54° to 90°. The Astronomers favored angles 45° to 30°. According to *The Secret Diaries of Hemiunu*", both groups of Craftsmen and Architects advised Sneferu to build his pyramid using their sacred angle. In order to appease both groups, Sneferu decided to use a sacred angle representing each group, and the result was the **"Bent Pyramid"**.

The "Bent Pyramid" whose proper name is "**The Southern Shining Pyramid**" was designed with its lower portion built at an angle of 54° to represent Ptah and the Primordial Mound, and its upper portion was built at an angle of 43° to represent Atum-RE (the Sun) and the sacred BenBen stone. Although some archeologist believe that the change in angle of the "Bent Pyramid" occurred due to building malfunctions, the actual diaries of the Architects who lived during that time indicate that the "Bent Pyramid" was intentionally designed to have a change in angle. More supporting evidence of the intended change in angle of the "Bent Pyramid" is found by observing that the angle of the "Red Pyramid" (whose proper name is "**The Northern Shining Pyramid**") is 43° and the angle of the smaller Pyramid that sits in front of the "Bent Pyramid" is 54°. The "Red Pyramid" and the smaller Pyramid were built to be representations of the **KA** or "Energy" of the upper part of the Bent Pyramid and the lower part of the Bent Pyramid symbolic of the Primordial Mound and BenBen stone respectively. So much like Dynastic Egypt was started by the Unification of Upper and Lower Egypt, Sneferu's "Bent Pyramid" represents a unification of Philosophies in the form of Construction. The advances made by Sneferu in Pyramid Building would be further advanced by his son the Pharaoh Khufu and his nephew the Architect Hemiunu.

Some hypothesis about how the Pyramids were constructed state that perhaps the blocks of the Pyramid were somehow "levitated" into position. Proponents of this philosophy point to the fact that the red color of Sneferu's "Red Pyramid" is due to high amounts of iron in the blocks, and this iron content could somehow make the blocks able to be levitated by magnetism. While the "levitation" Pyramid construction hypothesis is intriguing, analysis of Khufu's great Pyramid has yielded a more plausible method. Khufu's Great Pyramid of Giza, (whose proper name is Akhet-Khufu meaning Khufu's Horizon) is the largest and most massive of all the Ancient Egyptian pyramids. The Great Pyramid of Giza builds on the concepts and lessons learned from Sneferu, Huni, Imhotep, and Djoser. Investigation of Khufu's Great Pyramid of Giza indicates evidence of a **low-angled interior ramp spiraling around the Pyramid to the top**. While it is likely that the Granite Blocks of the pyramid were quarried and transported to the construction site, the use of this interior spiral ramp coupled with the ability to fabricate limestone blocks directly in place is the most plausible Pyramid construction theory proposed to date and demystifies the conundrum of how the Pyramids were built.

Khufu – Great Pyramid of Giza

Khufu's Great Pyramid at Giza, Khafre's Pyramid at Giza, and Menkaure's Pyramid at Giza represent the **Apex** of Pyramid building in Egypt. All of these Giza pyramids were built by the same blood-line of Pyramid Builders. Similar construction techniques and similar angles were used for the Giza Pyramids. The Great Pyramid was constructed with over 2 million blocks and finished with a polished limestone casing giving it a mirror-like quality which reflected the Sun. However, the polished limestone was stripped from the Pyramids by Arab invaders to build Mosques in Egypt. Egyptologists believe that the Pyramid Builders worked at a rate of placing 300 blocks per day. Some construction hypothesis suggest that the blocks were quarried and cut using acid or even Lasers, however, as discussed earlier, it is more likely that the majority of the blocks were fabricated and cast in place. Although Egyptologists are certain that the Step-Pyramid at Sakkara was a Tomb for a dead body, it is not certain that the Giza Pyramids were used as tombs for a dead Pharaoh due to the fact that no dead body or mummy has been found inside of their sarcophagus (**Neb-Ankh**). One explanation for the absence of a deceased mummy in the Pyramids is that the Pyramid was not intended to be a tomb for a dead person, but rather a figurative or symbolic tomb for a deceased deity, ideology, or principle.

Khafre – 2nd Largest Giza Pyramid

In Egyptian and African cosmologies, the concepts of "Life" and "Death" are often used symbolically to represent concepts beyond physical Life and physical death. Just like studying in the Library is called being "Buried in the Books", the knowledge and wisdom that goes into the development of a new Architectural structure is similarly and symbolically "Buried" within the creation of the structure. The "underworld" and "death" in Egyptian cosmology is often related to the rational world of mental processes. Consider that the type of environment that is conducive for Learning and Studying is also conducive for sleeping and burial.

Around the time that Sneferu was constructing his Pyramids, we see the first Trapezoidal, 1-layer Ziggurat built in Mesopotamia in the city of Uruk for the deity Anu. Since the various Egyptian structures pre-dated the Mesopotamian structures and due to the closeness of proximity between Mesopotamia and Egypt, it is our belief that at some point in time during the 4th Dynasty of Egypt, the Egyptians allowed Mesopotamians into Egypt to learn the science of building. The Egyptian Pharoah **Menkaure** was the last ruler of the 4th Dynasty to apply the Science of Pyramid Building before a major change occurred with the craft.

Menkaure – 3rd Giza Pyramid

The 5th and 6th Dynasties of Egypt are considered the "Decline" of Egyptian Pyramid building because it was during these dynasties that the method of Pyramid Building changed for the worse which is evident in the quality of the structures that remain. While most of the Pyramids of the 4th Dynasty are still standing, many of the smaller Pyramids of the 5th and 6th dynasties have crumbled. What remains of the 5th and 6th Dynasty Pyramids indicate that while the 4th Dynasty used what appears to be "cut stone" (or high quality fabricated limestone), the 5th and 6th Dynasty built pyramids using Mud bricks which have withered away over time. The 5th and 6th Dynasty Pyramids were also definitely used as tombs since mummies have been found inside their sarcophagus. The 5th and 6th Dynasty Pyramids are also the first time that we see the appearance of the "**Pyramid Texts**" which would eventually become known as "The **Book of the Dead**" or the "**Book of Coming Forth by Day**" (**Per-t-Em-Heru**). Notable Pyramid Builders of the 5th and 6th Dynasty were **Userkaf, Sahure, Unas, Teti**, and **Pepi II**. Strangely after the Pyramid Builders of the 6th Dynasty, Pyramid Building in Egypt becomes scarce to almost non-existent until the 12th Dynasty. Also, during the hiatus of Pyramid Building in Egypt is when we see the explosion of Ziggurat building in Mesopotamia.

5th Dynasty Pyramid of Unas
1st Appearance of the Pyramid Text

6th Dynasty Pyramid Complex of Pepi II

While the Egyptians built most of their structures as tombs, the Mesopotamians built Ziggurats as temples. Shortly after the Apex of Pyramid Building in Egypt, we see an explosion of Ziggurat building in Mesopotamia spear headed by Architect/Ruler **Ur-Nammu**. The first Mesopotamian Ziggurat was the 1-layer Trapezoidal **Temple of Anu** built around 2600 BC; almost 100 years after Imhotep designed Djoser's Step-Pyramid at Saqqara. The first true multi-tiered Ziggurats were built by the Mesopotamian Master Builder named **Ur-Nammu**. **Ur-Nammu** was responsible for the construction of the Ziggurat named **"ETEM-EN-NIGUR"** (House whose foundation causes terror) in the city of **UR** for the deity named **SIN**; the Ziggurat named **"EDUR-AN-KI"** (**House binding Heaven and Earth**) in the city of **NIPPUR** for the deity **ENLIL**; and the Ziggurat named **"EUNIR"** (House Temple-Tower) in the city of **ERIDU** for the deity **ENQI**. Considering that the word **"Ziggurat"** is phonetically similar to the word **"Saqqara"**, and considering that the first true stepped-pyramid shaped Ziggurat was built almost 700 years after the first step-pyramid at Saqqara, and the Mesopotamian builder of Ziggurats Ur-Nammu is always depicted wearing a "skull-cap" headdress similar to the Ancient Egyptian builders, it is a reasonable conclusion that Mesopotamian Ziggurat construction was inspired by Egyptian Pyramid Building.

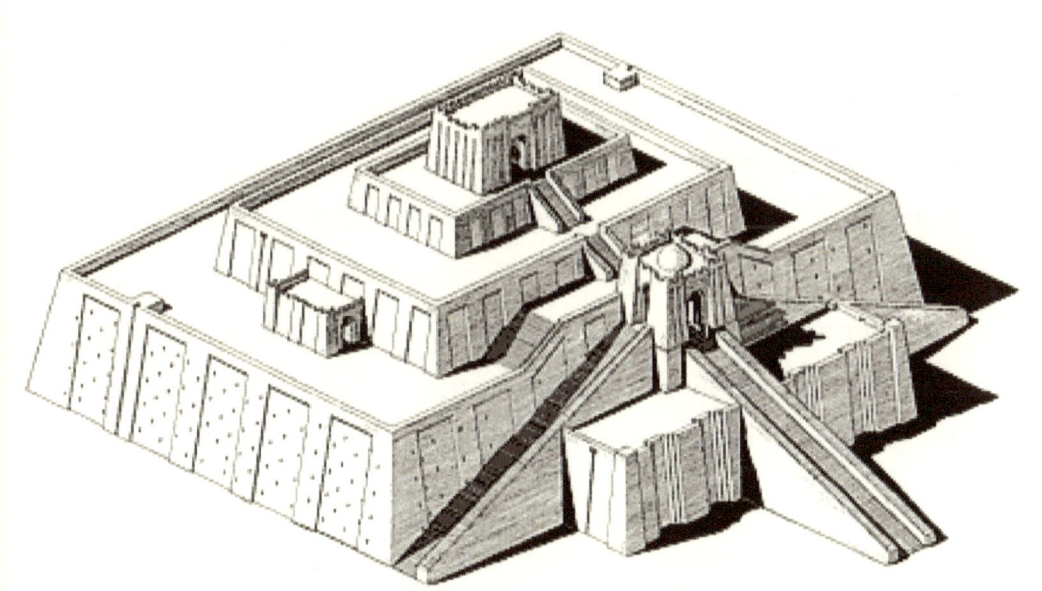

ETEMENNIGUR Ziggurat of UR for SIN

UR-NAMMU

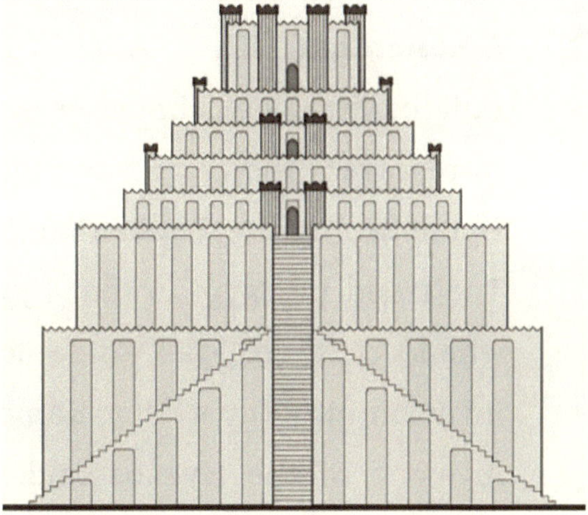

EDURANKI Ziggurat of NIPPUR for
ENLIL

UR-NAMMU Mesopotamian Master Builder of Ziggurats

After Pyramid Building had seemingly come to an end in Egypt after the 6th Dyansty, **Mentuhotep II** reunites Egypt during the 12th Dynasty. An attempt to resurrect the craft of Pyramid Building is initiated by the Nubian vizier **Amenemhat I** who becomes ruler after the death of Mentuhotep IV. While there was a definite and sincere effort to resurrect the craft of Pyramid Building during the 12th Dynasty, the Pyramid construction techniques of the 12th Dynasty resembled the construction techniques of the 5th and 6th Dynasty and did not quite meet the same quality achieved during the 4th Dyansty. The Nubian Vizier Amenemhet I initiated a bloodline of 12th Dynasty Pyramid Builders which included **Senusret I**, **Amenemhat II**, **Senusret II**, **Amenemhat III**, and **Amenemhat III**. After the 12th Dynasty, pyramid building in Egypt almost ceased again until another push to resurrect the craft was made during the 17th and 18th Dynasties by the **Thebes** family of **Kamose** and **Ahmose**. The trend in Egypt is that Pyramid Building would cease whenever there was some foreign influence or foreign invasion into Egypt. For Kamose and Ahmose, the foreign invasion of the Hyksos and the murder of their father **Seqenenre Tao II** marked the end of Pyramid Building in Egypt and the migration of the craft to other lands. Kamose and Ahmose would build a cenotaph (empty tomb) Pyramid as the last royal Egyptian Pyramid.

Step-Pyramid/Temple of Mentuhotep II

Pyramid of Amenemhat I – the Nubian Vizier who became a Ruler and resurrected the craft of Pyramid building during the 12th Dynasty

Pyramid of Senusret II "Senusret Appears" at Illahun (El-Lahun). Pyramid complex includes a satellite Queen's pyramid.

Built 2 pyramids: "Amenemhat III is Beautiful" aka "The Black Pyramid" and "Amenemhat III Lives"

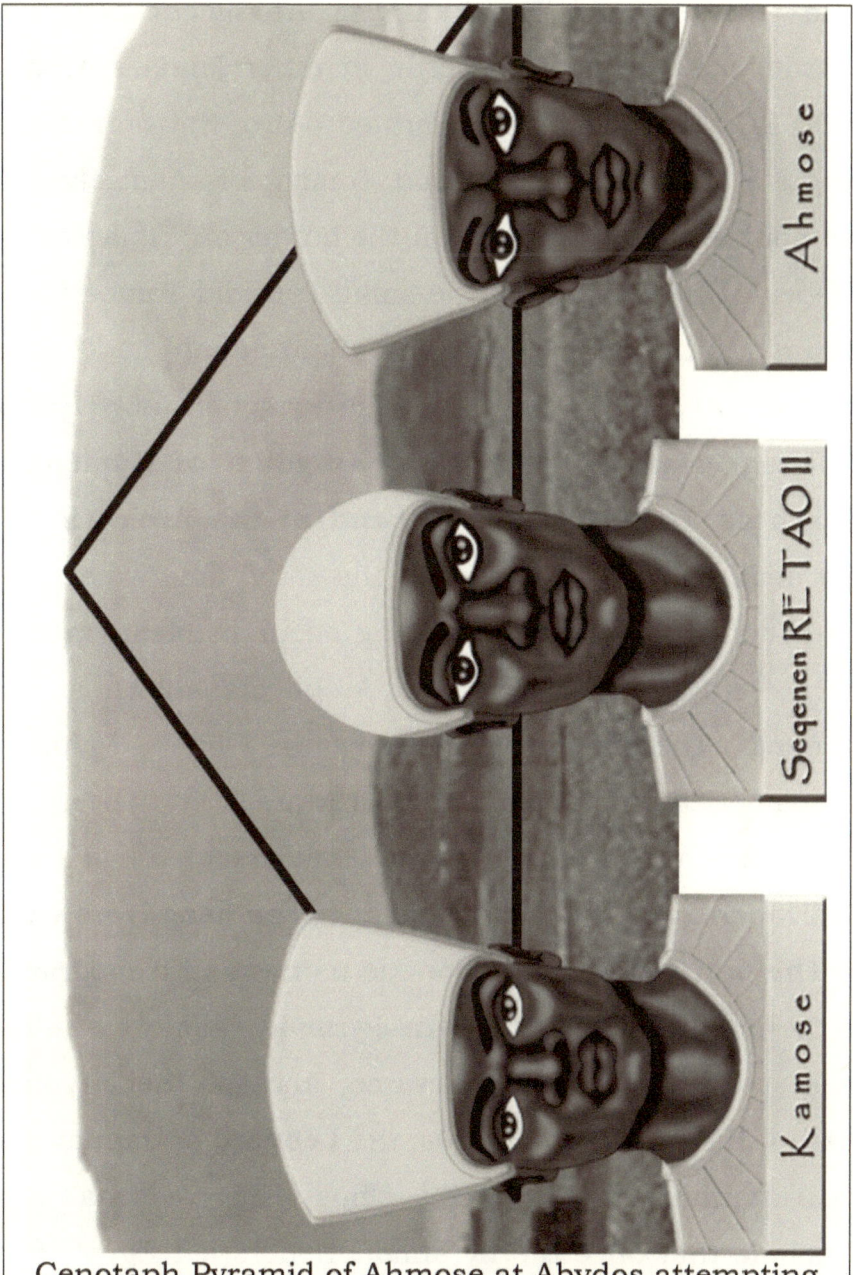

Cenotaph Pyramid of Ahmose at Abydos attempting
to resurrect the craft after the death of his father
Seqenenre Tao II

As the invasions and conflicts in Egypt increased over time, the Pyramid Builders and Architects began to migrate out of Egypt to settle in other lands. Around 1300 BC, Architects who migrated out of Northern Egypt began building small Pyramid Tombs for themselves in the Southern Egyptian city on the border of Nubian called Thebes. One example of the small Pyramid Tombs is the Pyramid built by the Architect **Sennedjem** in Thebes. Around 900 BC we see the resurrection of Pyramid Building occur in the **Nubian Kingdom of Kush** (also known as **Ta-Seti** meaning "**Land of the Arch**") in the city of **El-Kurru**. Also around the same time we see the resurrection of Pyramid Building occur in **Mesoamerica** with a group of Builders and Architects known as the **Olmecs**. The similarities between the Olmec, Egyptian, Mesoamerican, and Nubian cultures are great to overlook. The Headdress of the Pyramid Builders of all of these civilizations was the same (similar to **the headdress worn by the deity Ptah**); the phonetic names and iconography of the important deities of the Pyramid Builders of all of these civilizations was the same; and for the Egyptian Viziers, Nubians, and Olmecs, the **Leopard** or **Jaguar** had great symbolic significance. Thus, it is reasonable to conclude that once the Pyramid Builders left Egypt, they settled in the South at Nubia and in the West in Mesoamerica to resume the craft.

Pyramid Tomb of the Architect Sennedjem in Thebes

Pyramid Building in Nubia went through a similar growth, development, Apex, and decline cycle as it did in Egypt. Nubian Pyramid Building began in the city known as **El-Kurru**. The earliest structure built at El-Kurru was a Tumuli for the **Kandake** (a Nubian word for Queen) named **Makeda** around 1000 BC. The first Nubian Pyramid built at El-Kurru was for **Aserkamani** around 950 BC. The Nubian Pharaohs who would eventually conquer Egypt also built Pyramids at El-Kurru. Notable amongst the Nubian Pyramids built at El-Kurru are the Pyramids of the Pharaohs **Alara**, **Kashta**, **Piankhy**, **Shabaka**, **Shebitku**, **Taharka**, and **Tanutamuni**. It is a documented fact that the Nubians built the Pyramids of the Egyptian 4th Dynasty, and recall that the Philosophy of the Builders of the Pyramids centered on the deity Ptah, the Primordial Mound, and they favored Pyramid Angles steeper than 54°. Thus, it is no surprise that the Nubian Pyramids are characterized by Pyramid Angles between 60° and 75°. Nubian Pyramid Building in El-Kurru lasted from around 1000 BC to 600 BC. After 600 BC, Nubians began building pyramids in the city of **Nuri** until 300 BC. Around 300 BC, Nubian Pyramid Building reached its Apex in the cities of **Napata** (Gebel Barkal) and **Meroe**. Nubians built Pyramids in Meroe until 300 AD until the rise of Christianity which marked the end of the craft and a migration of Nubians into **West Africa**.

Alara – Nubian Pyramid El-Kurru 9

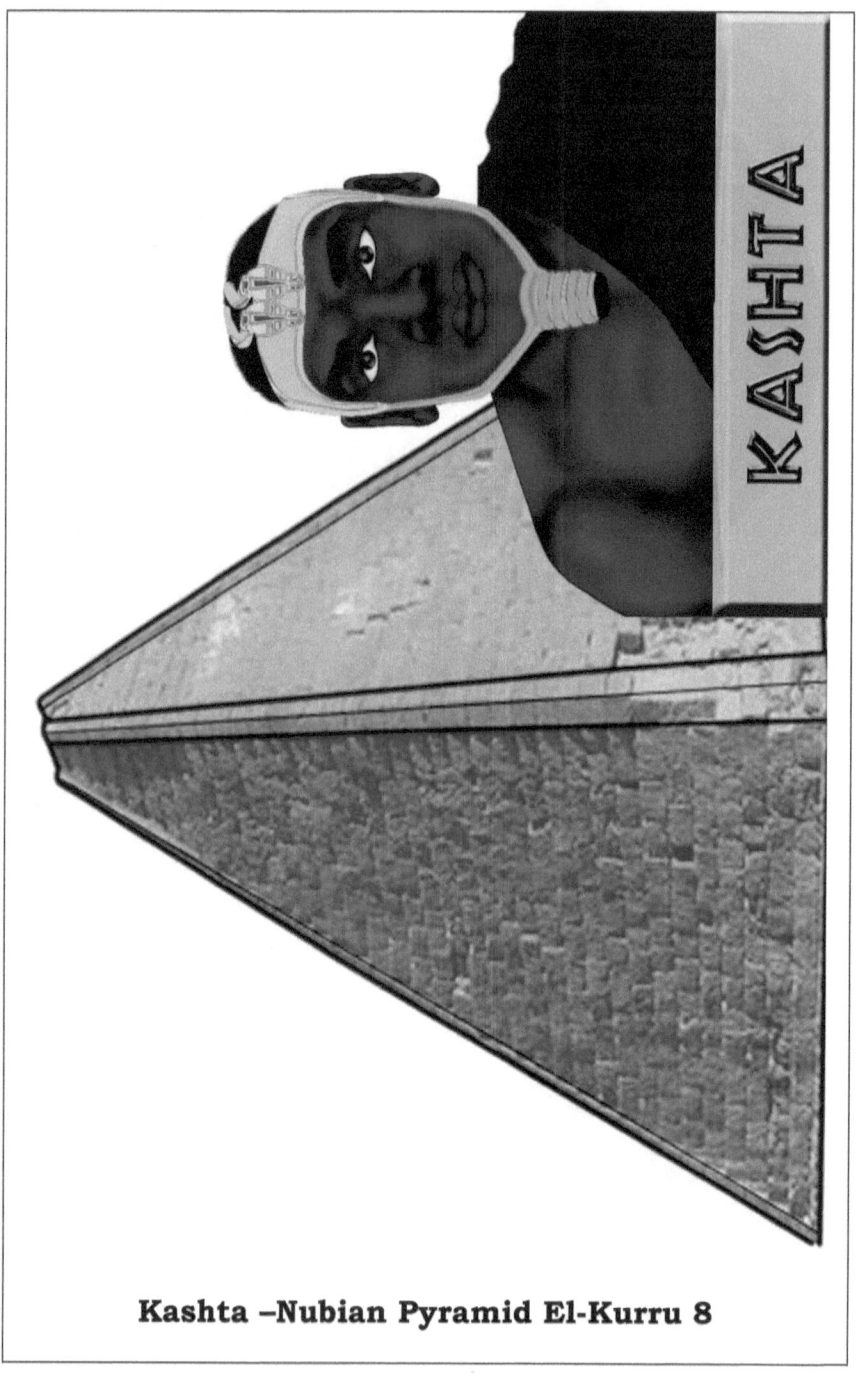

Kashta –Nubian Pyramid El-Kurru 8

Piankhy – Nubian Pyramid El-Kurru 17

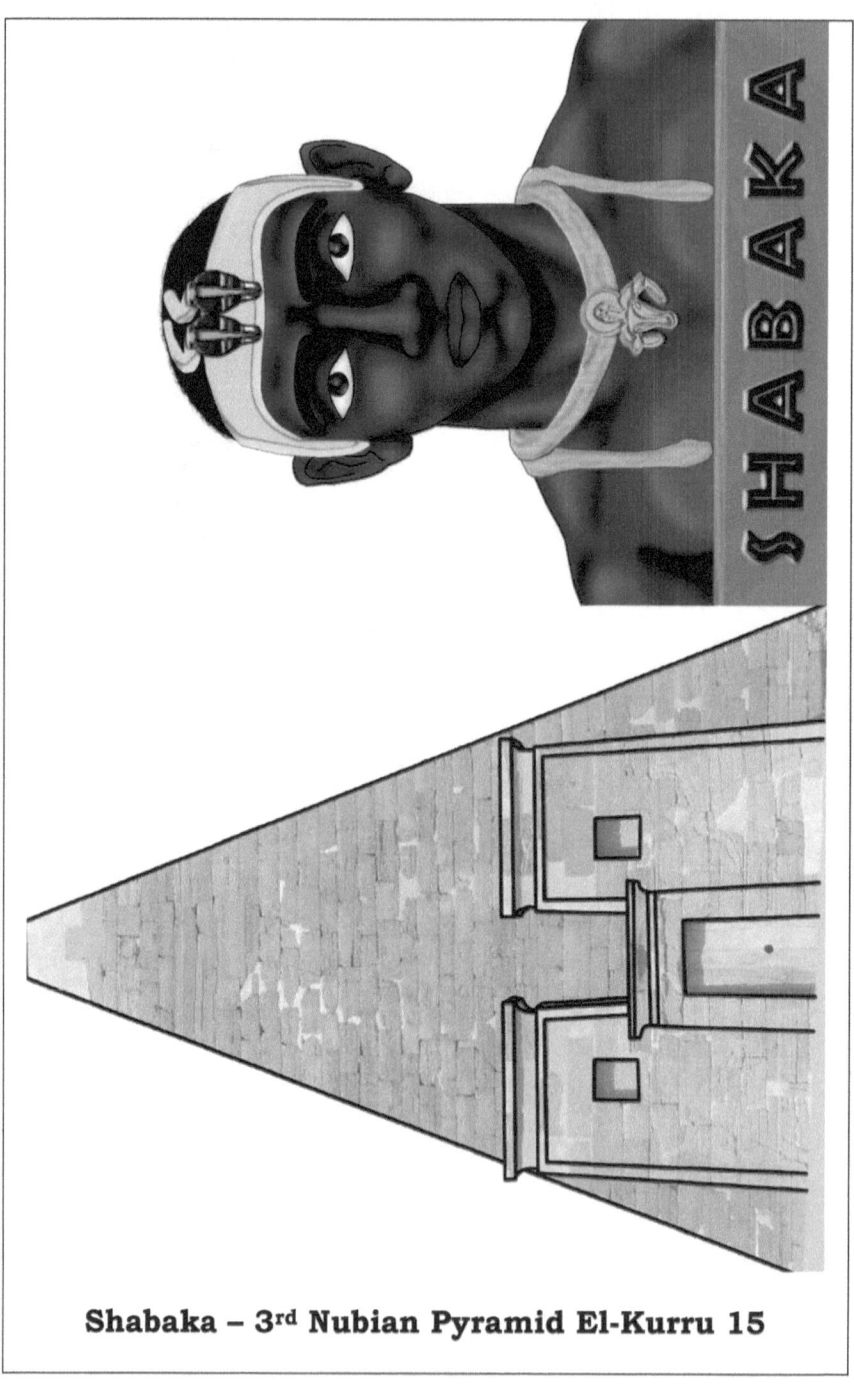

Shabaka – 3rd Nubian Pyramid El-Kurru 15

TaHarKa – Largest Nubian Pyramid

Kandake Amani-Tory – Nubian Pyramid (50 AD)

Nubian Pyramid building lasted more than 1300 Years (from 1000 BC to 300 AD). It is a fact that **there are more Pyramids in Nubia than anywhere else in the world,** and that most of the Pyramids build in Nubia were built for the Nubian Queens or **Kandake.** Thus, <u>**most of the Pyramids in the world have been built for Nubain Women**</u>.

While the craft of Pyramid Building was being resurrected in Nubia around 1000 BC, a group of **Traveling Men** known as the **Olmecs** brought the Science and Art of Pyramid Building to the Americas and settled in the area that is present day Mexico near the Yucatán Peninsula. While the Yucatán Peninsula was inhabited and populated prior to the arrival of the people known as the Olmecs, the arrival of the Olmecs in the area initiated an era of Pyramid Building which would last for 2400 years from 900 BC to 1500 AD. The Pyramid complex that is officially attributed to the Olmecs is the Pyramid Complex at La Venta, Mexico which features a step Pyramid separated by a plaza from The Great Pyramid of La Venta. The actual term "Olmec" was a name given to them by their descendants the Aztecs. The actual name the Olmecs called themselves is unknown, but there are speculations that the Olmec may have called themselves either **Tamoanchan** or the **Xi** (Pronounced Shi) people. There are 17 colossal Olmec stone heads that have been found throughout the Yucatán Peninsula of Mexico, and there are almost as many different Pyramid Complexes throughout the area as there are Olmec Heads. It is quite possible that each member of the traveling Olmec builders served as a patriarch of an eventual Pyramid building city in the Yucatán Peninsula.

The "Olmecs" – Traveling Students of the Craft who brought Pyramid Building to the Americas

The Olmec and Mesoamerican Pyramids were all built as "step-pyramids" and were used as Temples. As the Olmecs mixed and mingled with the other inhabitants of the Yucatán Peninsula, new tribes and cultures influence by the Olmecs would become learned in the Science and Art of Pyramid Building. The first Mesoamerican tribe to take up the craft was known as the **Toltecs**. The Toltecs were such renowned craftsmen that the word "Toltec" came to mean "**artisan**" in the Aztec Language. The Pyramid complex built at **Xochitecatl** which includes **The Pyramid of Flowers** and the round **Spiral Pyramid** (700 BC) and also the **Great Pyramid of Cholula** (300 BC) were built with dual Olmec/Toltec influence. The Toltecs are credited with building the Pyramids in Tula, Hidalgo and the Pyramid complex at **Teotihuacan** which includes The Great **Pyramid of the Sun**, The **Pyramid of The Moon** (called **Tenan**), and the **Temple of the Feathered Serpent** (100 AD). After the decline of the Toltecs came the rise of the **Mayans** who built the **Altun Ha Pyramids** (200 AD), The **Pyramids at Calakmul** (600 AD), The **Round Pyramid of the Magician** (700 AD), The **Tikal Pyramids and Temple of the Jaguar** (700 AD), and the **Pyramid Temple of Kukulkan** "the Feathered Serpent" (900 AD). Recall that in Egypt, "The Feathered Serpent" was a craft rode by the craftsmen deity Sokar, and in Mesoamerican culture this feathered serpent is called by various names including "Quetzalcoatl" and "Kukulcan". The Mayans have predicted the return of this **Craft** in the year **2012 AD**. As the Mayan culture declined, the Aztec culture would take up the craft of Pyramid Building. The Aztecs, who called themselves the **Nahua** People (the word Nahua means "intelligent" in the native language), were the last of the Mesoamerican Pyramid builders. The Nahua people built Pyramids at **Tenochtitlan** around 1300 AD which were eventually destroyed by Spanish Conquistadors in 1521 AD. Much like the invasion of the Christian converts of the Aksum Empire led to the destruction and decline of Pyramid Building in Kush and Nubia, the invasion of Spanish Conquistadors who destroyed Mesoamerican temples in the name of Christian ideologies led to the end of Pyramid Building in Mesoamerica. In order for balance to be obtained, Judeo-Christian structures must be destroyed in favor of traditional African ideologies and Architectural structures. The Mesoamerican culture appears to have been a blend between African and Asian cultures and ideologies. In fact many of the Mesoamerican and Nubian Pyramids are on the same lines of Latitude between 15°N and 19°N. Also, **the first emperor of China was buried under a massive Pyramid tomb in 200 BC**. In addition to Pyramids, the many mounds built throughout North America, and the mounds built in South and Central

America including the Yacatas at Tzintzuntzan, the Teuchitlan conic mounds, and the Cuicuilco Pyramid Mounds all seem to follow from this great building tradition. There are also Round Step Pyramid Mounds in **Guadalajara, Mexico** and in **Nigeria West Africa** called the **Nsude Arunsi Round Step Pyramid Mounds**.

As we follow the chronology of Pyramid Building, we see that the "Architects" or "Master Builders" in Ancient times were taught and trained in Ancient Egypt, and once the Apprentice student became a Master, they were then able to travel around the world Building, Working and Teaching the Craft of Pyramid Building. We see possible examples of this Lineage and Philosophy in the many Pyramids constructed around the world. The chronology of the Pyramids presented here is in the order generally accepted based on the information available to today's Archeologists. However, there are other alternative timelines and chronologies of Pyramid building that are not readily accepted by mainstream archeology. It is not the purpose of this book to speculate about the past, but rather to provide the reader with useful information on how to build a Pyramid in the present. In addition to the mainstream chronology, we will also present some of the alternative hypotheses of Pyramid chronology to give the reader a more holistic perspective of the subject. Most of the Pyramid Building chronologies place the origin of the craft in Africa in the present day countries known as Egypt and Sudan. Other alternative Pyramid Building chronologies place the origin of the craft with the "**Lost city of Atlantis**", and other hypotheses place the origin of the craft with **Extra-Terrestrials** from the planet **Mars** and the **Orion** and **Sirius** star constellations. One of the stronger alternative Hypothesis based on the empirical evidence of water erosion on the Sphinx and the obscurity and relative absence of any identifying glyphs on the Pyramids, places the dating of the **Har-Em-Akhet** (the Great Sphinx) to **10,500 BC**, and also states that the 3 Pyramids of Giza were built first, and all of the subsequent pyramids built in the area were done much later around 3000 BC in an attempt to copy and duplicate the already existing structures. Some people also believe that some curious structures found in the **Cydonia** region of the planet Mars are **5-sided Pyramids**, and a **Sphinx face**. These theories are also used to state an Extraterrestrial origin for Pyramid construction. Until we are able to actually travel to Mars and investigate these structures, we will not know for sure if they are "Pyramids" or Mountains. However, because we can track the development of pyramids from the Stepped-Pyramid, to the Bent Pyramid, to the Smooth sided pyramid, then if aliens did build pyramids on Mars, then come to Earth and built pyramids, the evidence shows that they somehow forgot the art and craft of Pyramid Building and had to re-develop it here on Earth.

Above: Suspected Pyramid Structure near a "Sphinx Face" in the Cydonia region of Mars

As we examine the geometry of modern religious Architectural structures we can clearly see the influence of the Ancient Nubian Domes and Pyramids. The Domed Buddhist **Stupa** Temples, the Pyramidal Buddhist **Borobudur** Temple, the **Spiral Ziggurats** in **Budapest** and **Samarra, Iraq**, the **Eskimo Igloos**, the Domes of Jewish Synagogues and Islamic Mosques, and the Steeples of Christian Churches all were derived from a tradition of Dome and Pyramid Building that started in Africa. Even the tallest Skyscraper buildings in the world are based on Pyramid Geometry. The **Burj Khalifa Building** in Dubai at 828 meters tall; the **Empire State Building** in New York City, United States at 381 meters tall; the **Ryugyong Hotel** in Pyongyang, North Korea at 330 meters tall; and the **Transamerica Pyramid Building** in San Francisco, California, United States at 260 meters tall are all some of the tallest buildings ever constructed which exemplify the stability of the Pyramid geometry in Architecture. And while there have been several modern constructions of large scale Pyramids such as the Glass Pyramid in front of the **Louvre Museum** in Paris, France or the **Pyramid Arena** in Memphis, Tennessee, one particular noteworthy modern Pyramid site was called **Tama-Re** or "**Egipt of the West**" built on over 400 acres of land near Eatonton, Georgia, U.S.A.

Above: Pyramids at Tama-Re built by the Nuwaubians

The Pyramids at Tama-Re were built in 1993 AD by a group called the **Nuwaubians** led by Dr. Malachi York **"Amun Nubi Re Ah Ptah"**. The Nuwaubians identified themselves as the descendants of the Ancient Mound Builders and Pyramid builders of Egypt, Nubia, and America. The Pyramid complex at Tama-Re marked perhaps the first time that the descendants of the Ancient Pyramid Builders have taken up the craft since Pyramid building ceased many years ago. When Malachi York was kidnapped and arrested in 2002 AD, the Tama-Re land was sold under government forfeiture, and the Pyramid structures were demolished by the local Georgia Sheriff's department and the FBI in 2005 AD. The Author and Architect of this book entitled **"ARCH I TECT: How to Build A Pyramid"** was fortunate enough to be able to visit the Pyramids on Tama-Re prior to their demolition and recognizes the monumental inspiration, motivation, and catalyzing force that building Pyramid Structures can have for a people. As a descendant of the Ancient Pyramid Builders, the Author and Architect of this book entitled **"ARCH I TECT: How to Build A Pyramid"** has provided all of the descendants of the Ancient Pyramid Builders a "Blueprint" or **"Black Print"** which will enable the ability to embark on the Science, Art, and Craft of Pyramid Building wherever they may travel in the world so that our structures will never be destroyed again.

BOREBORE

4.0. THE ARCH AND PYRAMID MATHEMATICS

The **Arch** is one of Nature's favorite geometric shapes when designing structures of great importance. Consider the Arch shape of your Rib cage used to protect your heart and other vital organs, or the Arch shape of Eggs used to house embryos of life. Since an "Egg" is a Structure which enables the development of life, and if we were to make an analogy to learning and education, then an Egg would be analogous to a school in that the school building is the structure in which mental growth and development takes place. Naturally, it would make sense to model physical school buildings after the symbolic "egg". In Architecture, Pyramids, Vaults, and Domes are good approximations of the "Arch" and "Egg" structure. Thus, it is no surprise that one of the many uses of Pyramids in Architecture was as centers of learning and education. After the Sphere, the Tetrahedron (Triangular Pyramid) is the Archetype and simplest Crystal structure constructed in 3-Dimensional space in Nature. The Tetrahedron (4 faces), Hexahedron (Cube, 6 faces), and Octahedron (Double Square Pyramid, 8 faces) all occur naturally in crystal structures, and also Bacteria, viruses, and the smallest elements of what we call life can be found configured in these Geometric formations.

The **Tetrahedron** (Triangular Pyramid) is considered the **Archetype** of Construction (meaning prototype and simple), the **Cube** is considered the "**Stereotype**" of construction (meaning often repeated), and the **Sphere** is the "**Epitome**" of Construction (meaning exemplified and greatest). The Sphere, The Tetrahedron (Triangular Pyramid), the Hexahedron (Cube), and the Octahedron (Double Square-Pyramid) can all be perfectly Blue-Printed in a symbolic pictogram called the "**Metatron Cube**". The Metatron Cube is derived from a more complex geometry which can be found on the walls of Egypt in the **Temple of Osiris** called the **Flower of Life** which is a symbol of Sacred Geometry used to depict the fundamental forms of Space in Nature and a template from which all forms come forth. The etymology of the word "**Metatron**" means "**behind the throne**" which alludes to the fact that the **Architects** and **Viziers** who

Flower of Life

Tetrahedron in the Metatron Cube

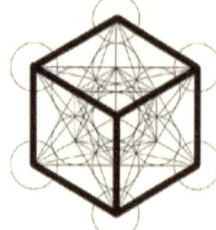

Hexahedron in the Metatron Cube

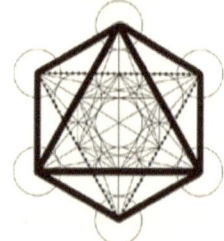

Octahedron in the Metatron Cube

were literally "behind the throne" in ancient times may

have had knowledge of the significance and application of the **Sacred Geometry** contained within the Metatron Cube. Connecting the edges of 4 Equilateral Triangles in the Metatron Cube creates the Tetrahedron (Triangle Pyramid). Connecting the edges of 6 Squares in the Metatron Cube creates the Hexahedron (Cube). Connecting the edges of 8 equilateral Triangles in the Metatron Cube creates the octahedron (Double Square Pyramid). The Sphere, Tetrahedron (Triangular Pyramid), Hexahedron (Cube), and Octahedron (Double Square Pyramid) were also used to symbolize the classical elements of Water, Fire, Earth, and Air respectively. Hence, one analysis of the word "Pyramid" comes out to be **"Pyro-"** and **"-Mid"** indicating **"Central Fire"**.

"Sacred geometry" is considered the Geometric forms, Angles, Ratios, Shapes, and other mathematic objects used in the design, planning, and construction of structures of great significant. To the Ancient Pyramid builders, one of aspects of Geometry that was considered Sacred was what we now know as the **"Angle of the Pyramid**." However, it is a fact that the Ancient Pyramid Builders did not measure Angles the way we measure angles today, but rather measured slopes using a method called the **"SEKED"**.

The SEKED was the Ancient Egyptian method for measuring the slope of an inclined surface. The SEKED of a Pyramid was calculated as the ratio of half of the Base of the Pyramid divided by the Height of the Pyramid. The calculation of the SEKED was equivalent to the modern Trigonometric calculation of the Cotangent of an angle as shown in the figure below:

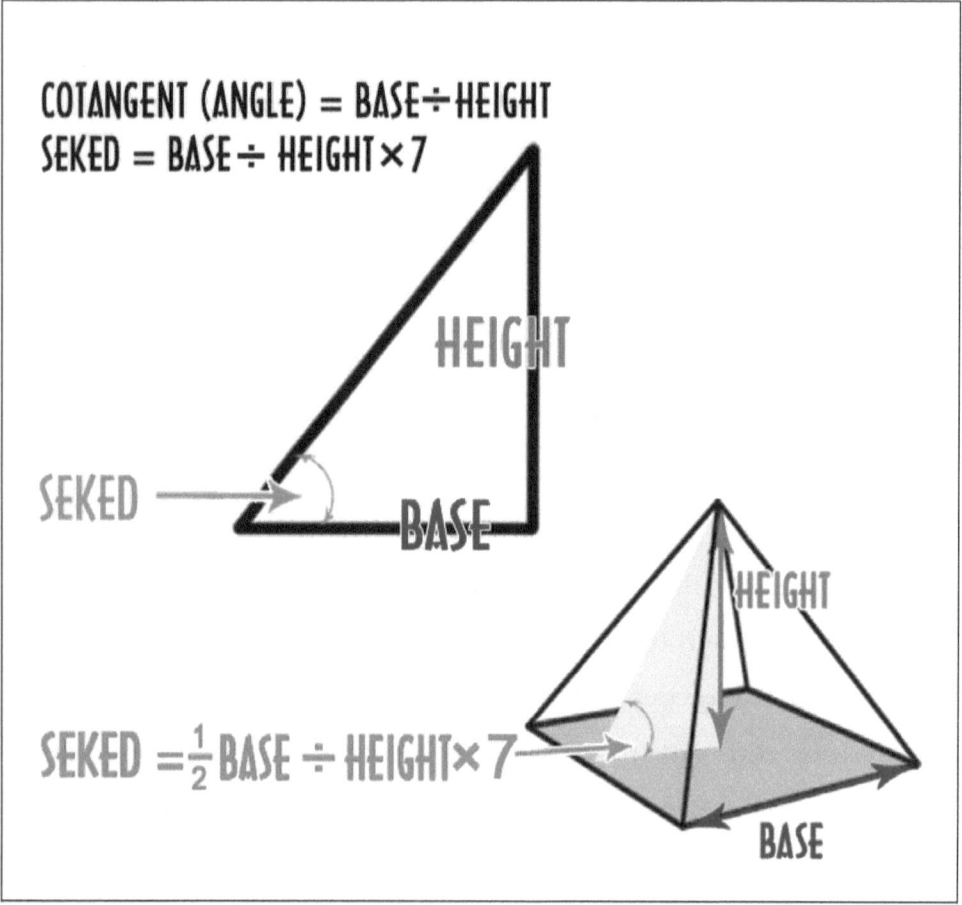

COTANGENT (ANGLE) = BASE ÷ HEIGHT

SEKED = BASE ÷ HEIGHT × 7

SEKED

HEIGHT

BASE

SEKED = ½ BASE ÷ HEIGHT × 7

HEIGHT

BASE

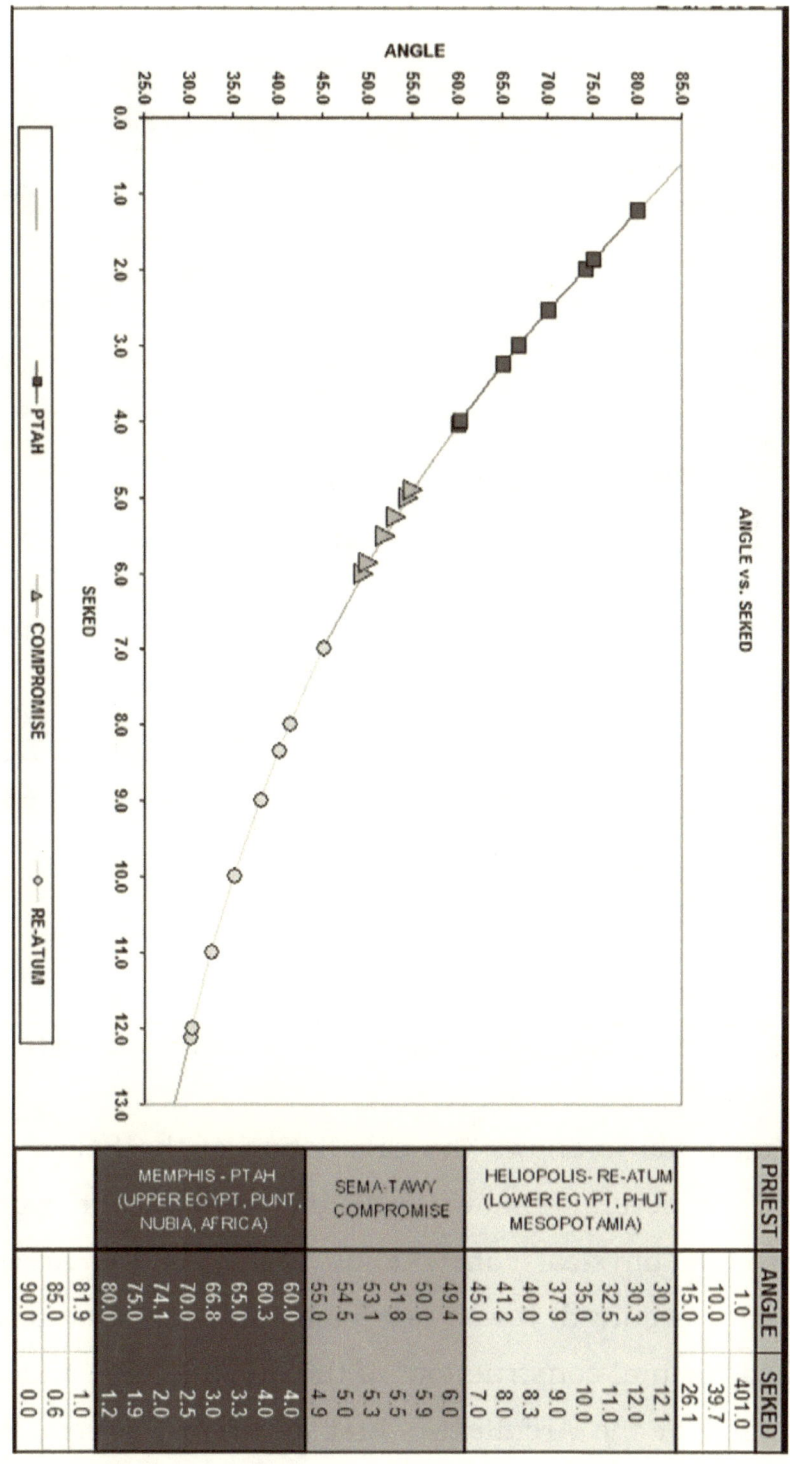

ANGLE vs. SEKED

PRIEST	ANGLE	SEKED
1.0	401.0	
10.0	39.7	
15.0	26.1	

	MEMPHIS - PTAH (UPPER EGYPT, PUNT, NUBIA, AFRICA)	SEMA-TAWY COMPROMISE	HELIOPOLIS- RE-ATUM (LOWER EGYPT, PHUT, MESOPOTAMIA)		ANGLE	SEKED
					30.0	12.1
					30.3	12.0
					32.5	11.0
					35.0	10.0
					37.9	9.0
					40.0	8.3
					41.2	8.0
					45.0	7.0
			49.4			6.0
			50.0			5.9
			51.8			5.5
			53.1			5.3
			54.5			5.0
			55.0			4.9
		60.0				4.0
		60.3				4.0
		65.0				3.3
		66.8				3.0
		70.0				2.5
		74.1				2.0
		75.0				1.9
		80.0				1.2
	81.9					1.0
	85.0					0.6
	90.0					0.0

Legend: ■ PTAH   △ COMPROMISE   ◇ RE-ATUM

SEKED

NAME	HEIGHT (meters)	BASE (meters)	ANGLE	SEKED
Pyramid Based on the Golden Ratio Triangle	-	-	72.0	2.3
Piankhy's Nubian Pyramid	10.0	8.0	68.2	2.8
Pyramid on the Back of the U.S. Dollar Bill	-	-	67.5	2.9
Mesoamerican Tikal Temple of the Jaguar	47.0	45.0	64.4	3.3
Taharka's Nubian Pyramid	50.0	51.8	62.6	3.6
Khafre's Pyramid at Giza	143.5	215.3	53.1	5.3
Khufu's Great Pyramid of Giza	146.6	230.3	51.8	5.5
Myan Step-Pyramid Temple of Kukulkan	30.0	55.3	47.3	6.5
Imhotep's Step-Pyramid for Djoser	62.0	124.0	45.0	7.0
Mesoamerican Step-Pyramid of the Sun	73.2	230.3	32.5	11.0

Table of Height, Base, Angle, and Seked of several Pyramids from around the world

The table above shows the calculation of the Seked and Angle for several famous pyramids around the world. Before the physical construction of any of the Ancient Pyramids occurred, it was extremely important that a well formed "mental construction" of the Pyramid take place in the mind of the Architect. Just as bricks, stones, and hammers are the materials and tools used to construct the Physical Pyramids, Mathematic calculations and equations are the tools required to construct the Mental Pyramid. The mental planning of a project using mathematics and the physical creation of that project is an example of the application of the Ancient maxim **"As Above, So Below"** and is the what is meant when it is said that the Pyramids were built from the "top, down".

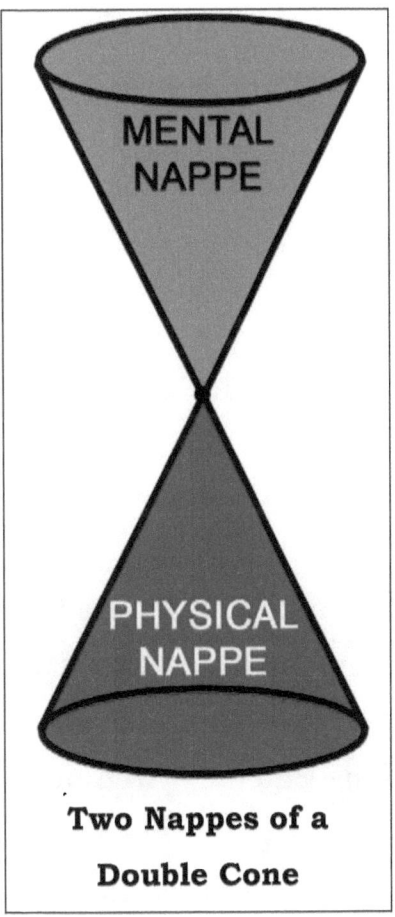

Two Nappes of a Double Cone

In mathematics, when two cones are placed apex to apex it is termed a **Nappe**, and the imagery of two cones placed apex to apex provides an illustration of an example of a Mental Pyramid built with Mathematics being reflected into a Physical Pyramid built with tools. Thus, the Mathematic techniques required to design a Pyramid will be discussed prior to the discussion about the physical construction of a Pyramid.

4.1. Catenary Arch

The U-shape that is formed by a chain hanging at its ends between two poles is called a "**Catenary Arch'**. The word "Catenary" means "**chain**" in Latin, and in the Akan culture of West Africa the word for "chain" is **Abosam**, which is also a word used for **Deities**. Thus, in the Twi Language of West Africa, the Catenary Arch can be called the "**Arch of the Abosam**". The Catenary Arch is expressed by the Mathematical equation:

$$y = \left(\frac{a}{2}\right) \times \left(e^{\left(\frac{x}{a}\right)} + e^{\left(\frac{-x}{a}\right)}\right) \qquad \text{where:}$$

> **a** determines the rate the arch "opens up"
>
> **y** is the height of the arch, and
>
> **x** is the width of the arch

Because a chain is composed of separate links, the Catenary Arch form can be constructed using rectangular bricks where each **brick is like a link in the chain**. While the U-shape of the graph of the Catenary Arch may resemble the more familiar algebraic equation known as the **Parabola**, it is significantly different. The Parabola is characterized by the mathematic equation **y=x^2**, and the word "Parabola" comes from the word **"Parable"** which is a **"saying or story where something is expressed in terms of another thing**." However, the mathematic form of the Catenary Arch is considered at **"Transcendental Function"** based on the **Napier** constant (**e = 2.718…**), and of course the word "Transcendental" means **"to go beyond"**.

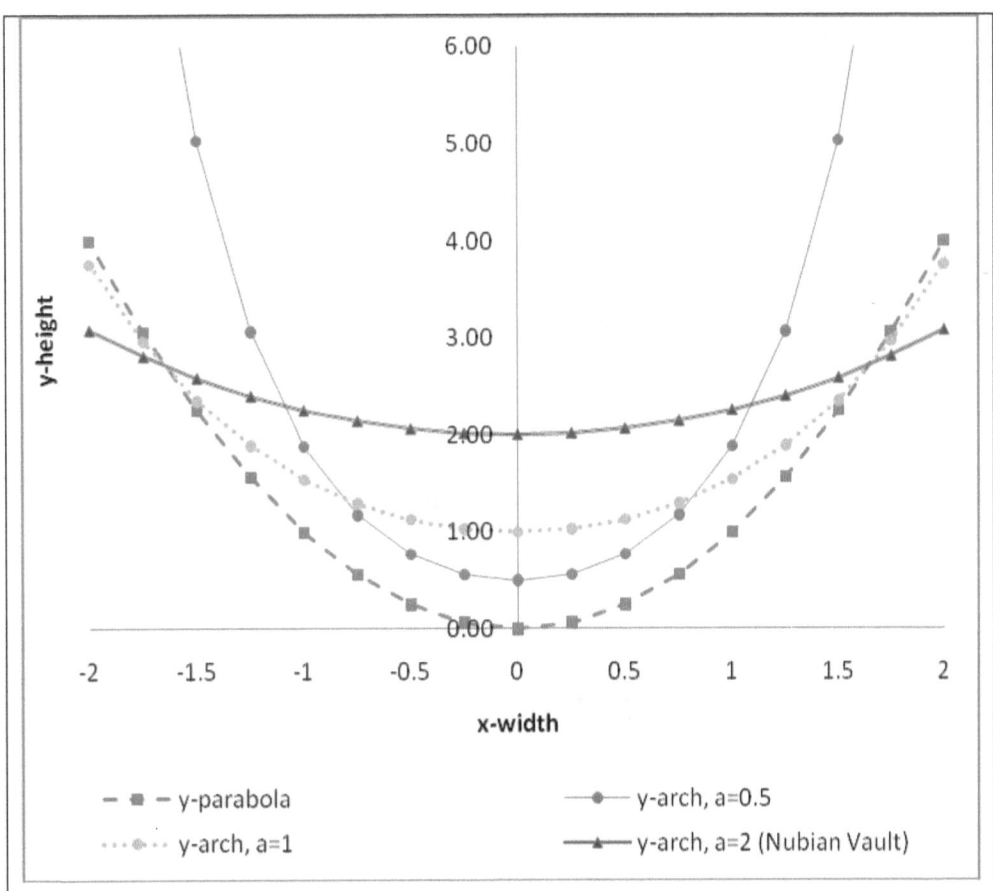

Above: Graph of the Catenary Arch for different values

An **Inverted Catenary Arch** can be constructed in brick masonry and can also be used to approximate the shape of a Pyramid. Vaults and Arches can also be approximated in brick masonry by using a process called the **Corbel Arch** which gradually "staggers" or offsets each course of bricks to form a "false Arch" or vault. The graph on the next page shows the relationship between an Inverted Catenary Arch, an Inverted Parabola, and the Piecewise linear function used to represent a pyramid characterized by the equations:

Inverted Catenary Arch: $y = \left(\dfrac{-25}{2}\right) \times \left(e^{\left(\frac{x}{-25}\right)} + e^{\left(\frac{-x}{-25}\right)}\right)$

Inverted Parabola: $y = \left(-x^2 + 2916\right) \times \left(\dfrac{75}{2916}\right)$

Piecewise "Pyramid" linear function:

For $-63 \leq x \leq 0$: $\quad y = \left(\dfrac{100}{63}\right)x + 100$

For $0 \leq x \leq 63$: $\quad y = \left(\dfrac{-100}{63}\right)x + 100$

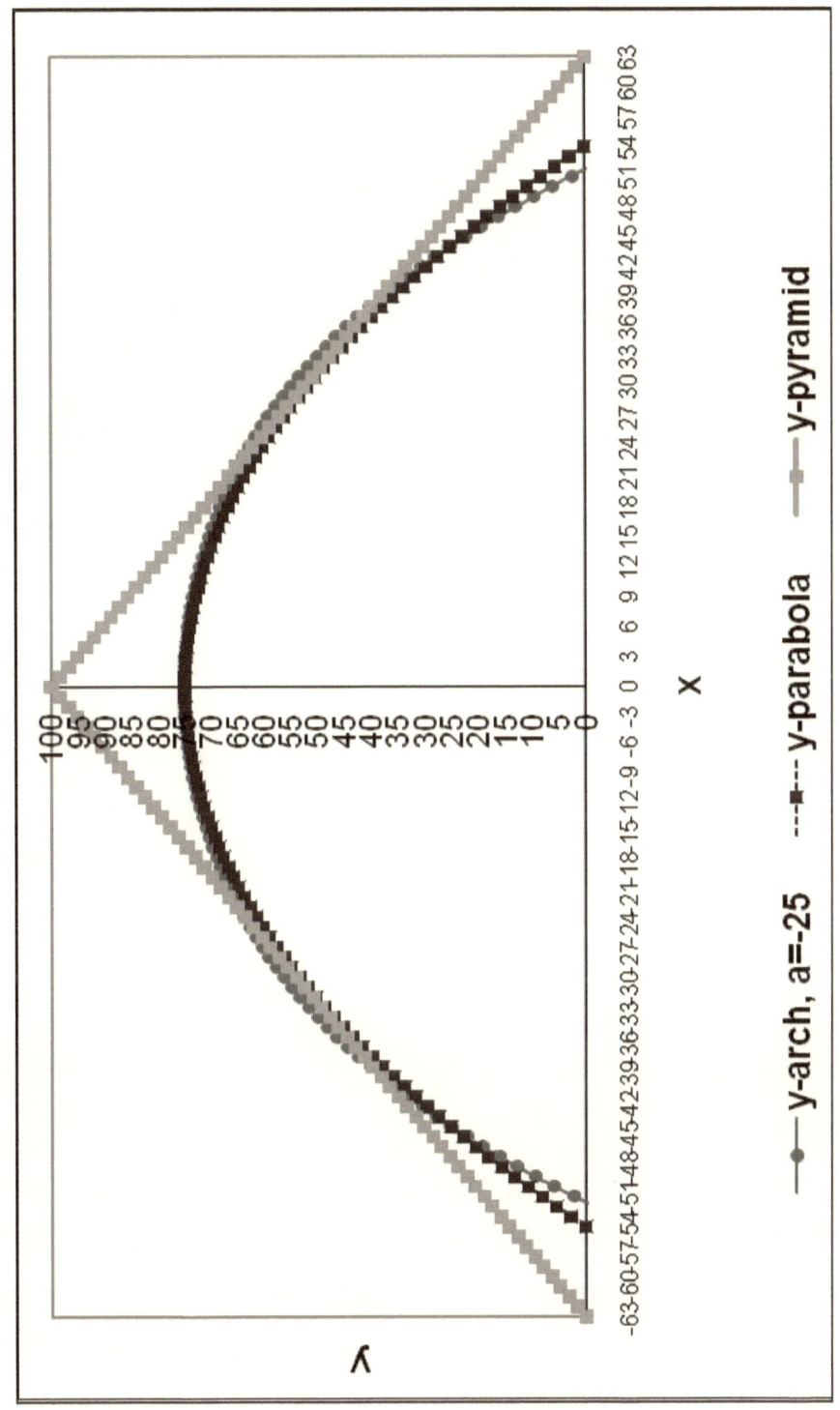

x

y

y-arch, a=-25 ----- y-parabola ---- y-pyramid

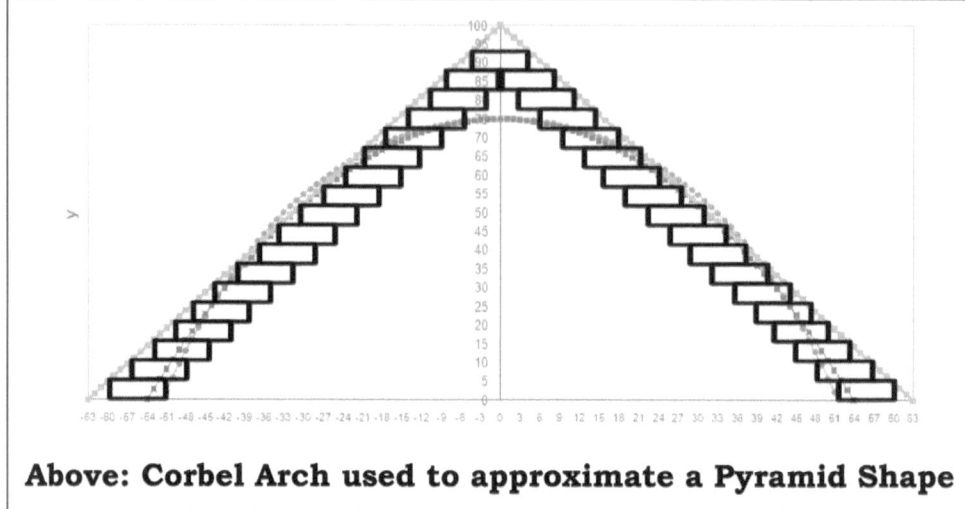

Above: Corbel Arch used to approximate a Pyramid Shape

The Mathematics of the Corbel Arch will be further explored in the section on Trigonometry.

Arches and Vaults provide us with a basic 2-dimensional method to approximate

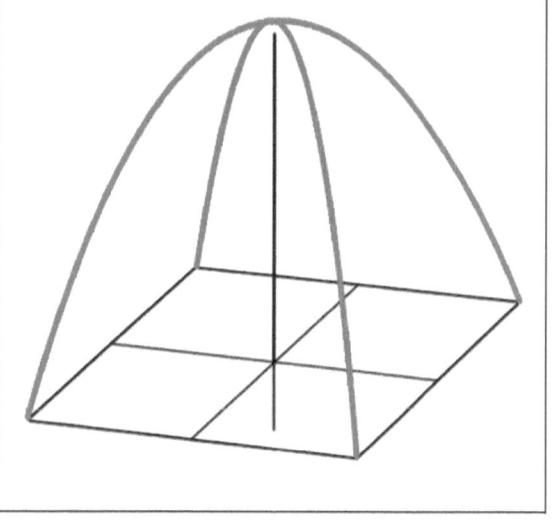

triangles. By crossing two Arches in 3-Dimensions we will be able to approximate the Pyramid shape. The portion of the Triangle (or Pyramid) that is not included in the Arch can be constructed as a capstone with the same angle as the overall Pyramid.

4.2. Nubian Vaults and Domes

In Mathematics, the **Nubian Vault** is a specialized version of the Catenary Arch where a=2 given by the equation:

$$y = e^{\left(\frac{x}{2}\right)} + e^{\left(\frac{-x}{2}\right)}$$

The Nubian Vault is a 2-dimensional Arch, and a 3-Dimensional **Nubian Dome** can be created by crossing two Nubian Vaults at a right angle. When the Nubian Dome is placed on a Square plane, the resulting form is called a **Cloister Vault** or **Domical Vault** in Architecture. In Architecture, the word "cloister" means "enclosure" and refers to a rectangular open space enclosed by walls. In West African Architecture, the Cloister is called **FIHANKRA** and is symbolized by the Adinkra symbol:

Thus, the Nubian Dome, Cloister Vault, and Domical Vault can be considered a **Fihankra Dome** or **Fihankara Vault** in the Twi Language of West Africa.

Another method of constructing a Nubian Dome on a Square Foundation is called a **Squinch**. The squinch was developed in Architecture as a means to **"Place a Circle (Dome) on a Square (Foundation)"**. Squinches are created by building successive courses of diagonally placed Nubian Vaults from the right-angled corners of a square base. The method of building Nubian Domes using the squinch Nubian Vault technique is common in South and Central American Architecture and is known as **Bovedas** or **Bobedas**. The figure below shows how a Pyramid can be constructed using a Pointed Nubian Vault Dome on Squinches:

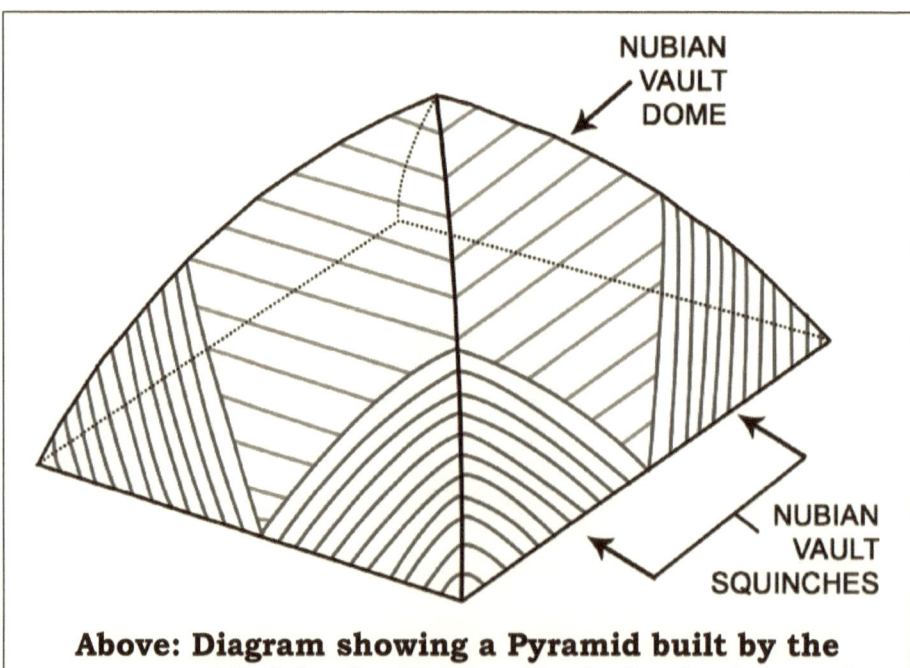

Above: Diagram showing a Pyramid built by the courses of Bricks for Nubian Vault Squinches and topped with a Nubian Vault Dome

Constructing a Pyramid using the Pointed Nubian Vault Dome on Squinches technique should also include a tie beam. Construction techniques for building Nubian Vaults and Nubian Domes can be found in a book entitled **"Building with Earth"** on pages 124-126 where it states:

*"Nubian Vaults can be built without the use of formwork by reclining the bricks to create arches. The shape of the bricks used in Nubian Vaults should have a low brick weight per unit area of mortar to prevent the bricks from sliding during construction. The degree of inclination of the arches is usually between **65°** and **70°** with the horizontal. Nubian Vaults only need two vertical walls onto which the inclined arches lean. It is also possible to lean the arches against a central "supporting arch". The shape of the vault is controlled by **"stretching a cord"**. Wedging a small stone chip on the outer edge of the arch brick helps to display arch action before the mortar is dry. The Nubian dome technique creates spherical domes. The Nubian dome technique is created by laying circumferential courses of bricks using a movable guide. With the Nubian Dome Technique, blocks are turned on their edge to avoid slippage of the freshly laid blocks. When spanning large areas, Nubian domes can be strengthened by reinforcing the dome with steel strips or concrete ring beams."*

The Mathematic equations for Volume allow us to know how much space is contained within our structures. The equations for Volume of a Square Pyramid and Spherical Dome are presented below:

Volume of a Square Pyramid:

$$volume = \frac{1}{3} length \times width \times height$$

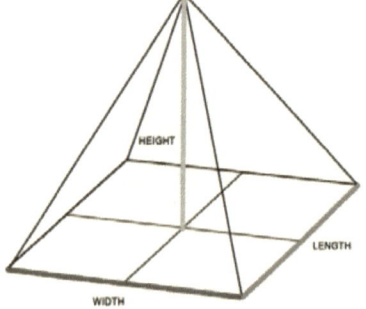

Volume of a Spherical Dome:

$$volume = \frac{4}{6} \times \pi \times r^3$$

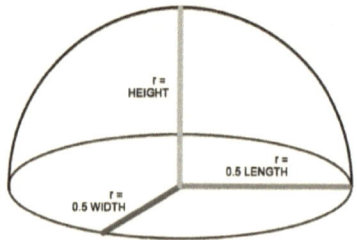

The Science of Nubian Domes can be further expanded and applied to the Science of **"Building Planets"**. Modern examples of model "Planet Building" include the **"Spaceship Earth"** building at the EPCOT (Experimental Prototype Community of Tomorrow) Center in Orlando, Florida, USA, **Biosphère 1** in Montreal, Canada, and **Biosphere 2** in Oracle, Arizona, USA. All of these "Planet Building" projects have been based on the geodesic dome design that is similar to the **"Ley Lines"** that are said to exist in the construction of planet Earth.

4.3. Trigonometry and Calculus

Ever since the construction of Imhotep's Step Pyramid for King Djoser, all of the Pyramids constructed with block, brick, or stone have utilized successive offset courses of blocks in order to form the angle of the Pyramid. In construction and Architecture, the process of gradually offsetting each successive course of blocks is called **Corbelling**. The word "corbel" comes from the Latin word for a "Crow" or "Raven" and is used because one offset brick on top of another has the appearance of a bird's beak. In the Twi language of West Africa, we would call the process of "Corbelling" blocks **"Obereku"** meaning "bird's beak". We can utilize the mathematics of Trigonometry to calculate the angle of the Pyramid created by our successive courses of offset blocks. When we do not offset the blocks, we obviously construct structures at 90° degree right angles. When we offset the blocks at a distance that is less than or equal to the height of the blocks, then we will construct structures between 45° degrees and 90° degrees. When we offset the blocks at a distance greater than the height of the blocks, then we will construct structures less than 45° degrees.

HEIGHT-LENGTH SECTION OF CORBELLED BRICKS	HEIGHT-WIDTH SECTION OF CORBELLED BRICKS
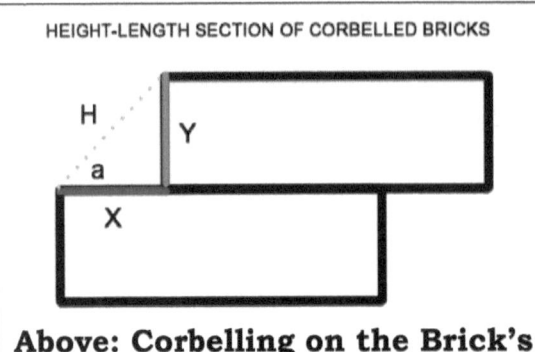	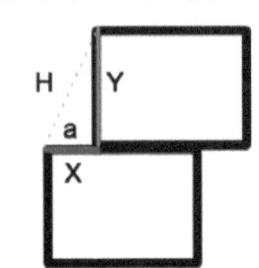
Above: Corbelling on the Brick's Length	**Above: Corbelling on the Brick's Width**

In order to ensure stability, the distance X that a brick is staggered or corbelled should not exceed $\frac{1}{3}$. In the figures above, if corbelling on the Brick's Length then $X \leq \frac{1}{3} \times Length$, and if corbelling on the Brick's Width then $X \leq \frac{1}{3} \times Width$. The angle created by corbelling the block is a function of the height of the block given by the table:

X (Offset Distance)	ANGLE
0	90
0.32×HEIGHT	72
0.58×HEIGHT	60
0.75×HEIGHT	53.1
0.79×HEIGHT	51.8
HEIGHT	45
1.73×HEIGHT	30

All of the Ancient Pyramids were built using the corbelling method. The blocks on the pyramids are all different sizes and shapes with the blocks on the bottom being the largest and getting smaller towards the top. The use of heterogeneous blocks (different size and shaped blocks) insures the sturdiness and stability of massive structures against Natural disasters. The mathematics has to have been precisely calculated for each block to maintain the same corbelling distance and maintain the same slope to the apex of the Pyramid.

To create smooth sided blocks for a Pyramid built using the corbelling method, we can use the Right Triangle Rule to calculate the **"Hypotenuse"** or diagonal side of each block used to create smooth sides. The length of the Hypotenuse is given by Right Triangle Rule where X is the corbelling offset distance and Y is the height of the block:

$$H = \sqrt{X^2 + Y^2}$$

The etymology of the word "Hypotenuse" means **"to stretch under"** and hence is related to the Ancient Egyptian Goddess **Sheshat** and the **"stretching the cord"** Pyramid building ritual. The relationship between the corbelling offset distance (X), the Height of the block (Y), the Hypotenuse (H), and the angle (a) can be determined using the following Trigonometric Equations:

$SIN\ a = \dfrac{Y}{H}$	$COS\ a = \dfrac{X}{H}$	$TAN\ a = \dfrac{Y}{X}$

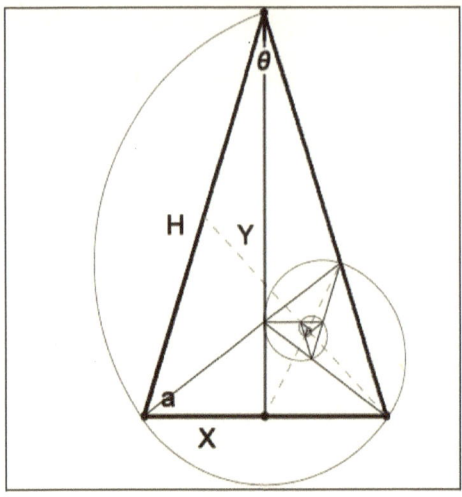

The proportions of a Pyramid based on the **"Golden Triangle"** also called the **"Sublime Triangle"** are given by the Triangle on the right, when the angle **a=72°** and the ratio of the corbelling offset distance (X) and the Hypotenuse (H) are equal to: $\dfrac{H + 2X}{H} = \dfrac{H}{2X}$. Recall that Pyramids built using the Golden Triangle proportions and angles in the 72° degree range were common amongst the Nubian Pyramid Builders of Kush.

Using the corbelling offset method from a square base will obviously create a Square Pyramid. The corbelling method can also be used from different shaped bases including triangular, pentagonal, etc, however, the most stability is achieved when the corbelling offset method described here is used from a circular base. Using the corbelling offset method from a circular base will create a Cone Pyramid which is stable enough to be scaled up as long as the proportions described here are maintained. The volume of the Cone Pyramid constructed using the corbelling offset method from a circular base can be calculated using the equation:

Volume of a Cone:

$$volume = \frac{\pi \times width \times length \times height}{12}$$

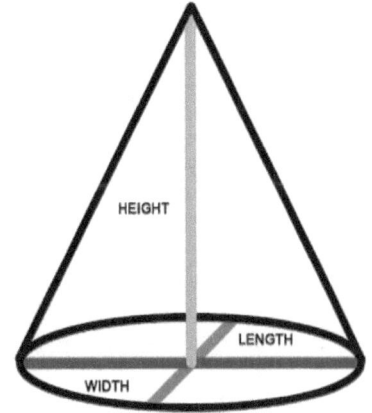

The mathematics of Trigonometry is an important tool for our corbelling building technique and it can also be used to build the frames and forms for our Nubian Vaults, Arches, and Domes.

Suppose we would like to build a 60° degree diagonal arm for the frame of a Nubian Vault given by the equation:

$$y = \left(\frac{-25}{2}\right) \times \left(e^{\left(\frac{x}{-25}\right)} + e^{\left(\frac{-x}{-25}\right)}\right) + 100$$

which can be approximated by an inverted parabola given

by the equation: $y = \left(\frac{-75}{2916}\right)x^2 + 75$

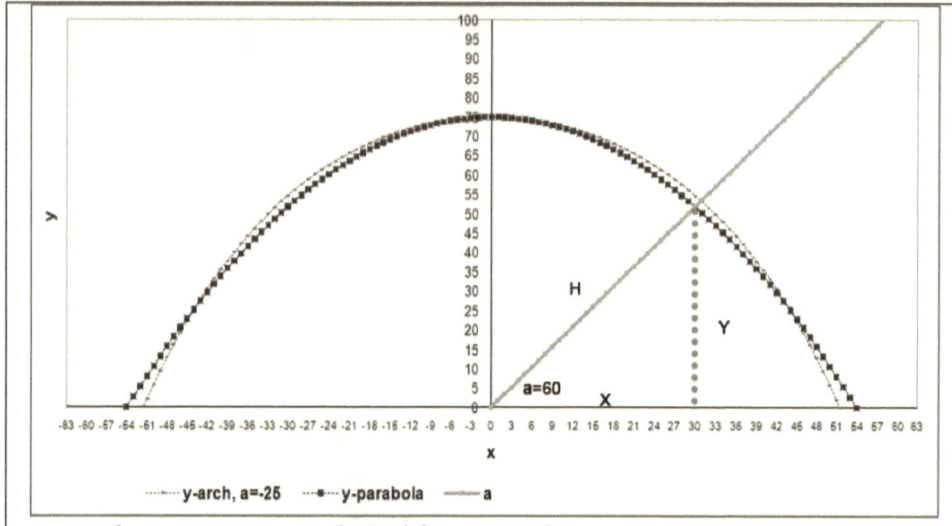

Above: Inverted Nubian Vault approximated by a Parabola with a 60° degree diagonal frame arm

We can use the TAN trigonometric function to determine the slope of the equation for the line of the 60° degree diagonal frame arm which is:

$$Y = TAN(60°)x = 1.732x$$

To determine the point on the Parabola where the equation for the line of the 60° degree diagonal frame arm intersects, we set both equations equal to each other and solve for x such that:

$$1.732x = \left(\frac{-75}{2916}\right)x^2 + 75 \text{ and}$$

$$\left(\frac{75}{2916}\right)x^2 + 1.732x - 75 = 0$$

We solve equations in the form $ax^2 + bx + c = 0$ using

the quadratic formula $x = \dfrac{-b \pm \sqrt{b^2 - 4ac}}{2a}$ such that

$$x = \frac{-1.732 \pm \sqrt{1.732^2 - 4 \times (75/2916) \times (-75)}}{2 \times (75/2916)}$$

Thus, x=29.96697. We can then solve for the length of the 60° degree diagonal frame arm H using our COS trigonometric equation such that:

$$H = \frac{X}{COS(a)} = \frac{29.96697}{COS(60)} = 59.9326$$

We can then use these equations to calculate a table of diagonal frame arms under our Parabolic Arch.

Diagonal Lines intersecting with the Inverted Parabola given by the equation y=(-75/2916)*(x^2)+75						
Diagonal Frame Arm (degrees)	a	b	c	X	Y	H
0	0.026	0.00	-75	54.00	0.00	54.00
15	0.026	0.27	-75	49.04	13.14	50.77
30	0.026	0.58	-75	43.93	25.36	50.73
45	0.026	1.00	-75	37.95	37.95	53.67
60	0.026	1.73	-75	29.97	51.90	59.93
75	0.026	3.73	-75	17.89	66.77	69.12
80	0.026	5.67	-75	12.51	70.97	72.07

The mathematics of Calculus uses multiple line segments to approximate a curve. When we build an Arch or Vault out of bricks, we are similarly using line segments (bricks) to approximate a curve (Arch). In order to mathematically determine the overall length of our Arches or Vaults, we must use another one of our mathematical tools from Calculus to calculate the Arc length. The Length of an Arc is given by the integral equation: $s = \int_{\alpha}^{\beta} \sqrt{1 + \left(\dfrac{dy}{dx}\right)^2}\, dx$

For a Parabola given by the equation $y = ax^2 + b$, we can evaluate the above integral to determine the arc length S from the center of the Parabola to the half-width β such that:

$$S = \frac{\beta}{2}\sqrt{1 + (2a\beta)^2} + \frac{1}{4a} LN\left(2a\beta + \sqrt{1 + (2a\beta)^2}\right)$$

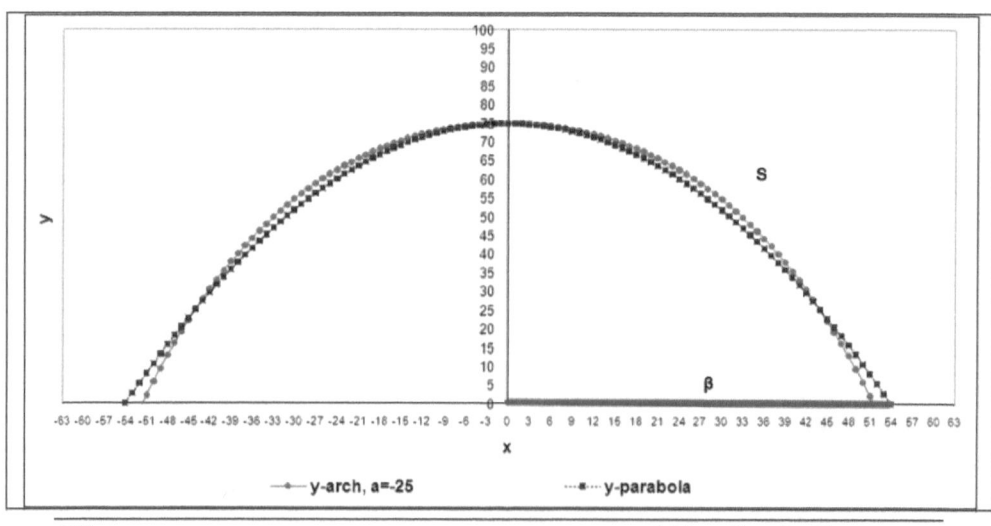

We can then multiply S by 2 to get the length of the Parabola from end-to-end. For a parabola given by the equation $y = \left(\dfrac{-75}{2916}\right)x^2 + 75$ which has a half-width

β=53, we can calculate the Arc Length S=96.6, thus the Arc Length of the Parabola from end-to-end is 193.2.

The Arc length (S) of a Catenary Arch given by the equation $y = \left(\dfrac{a}{2}\right) \times \left(e^{\left(\frac{x}{a}\right)} + e^{\left(\frac{-x}{a}\right)} \right)$ from the center of

the Catenary Arch to the half-width β is:

$$S = a \times \left(e^{\left(\frac{\beta}{a}\right)} - e^{\left(\frac{-\beta}{a}\right)} \right)$$

For a Catenary Arch given by the equation

$$y = \left(\dfrac{-25}{2}\right) \times \left(e^{\left(\frac{x}{-25}\right)} + e^{\left(\frac{-x}{-25}\right)} \right) + 100 \text{ which has a}$$

half-width β=51.5859, we can calculate the Arc Length S=96.8, thus the Arc Length of the Catenary Arch from end-to-end is 193.6. So we can see how closely the parabolic equation approximates the Catenary Arch.

5.0. THE PYRAMID SCHEMES

In modern terms a **"Pyramid Scheme"** is a term used to describe selling individuals "hopes and dreams" by encouraging more and more people to participate in the "scheme" without ever really providing any real solutions, products, goods, and/or services. However, the concepts presented here in this book are not part of a "Pyramid Scheme" in the derogatory sense of the phrase, but rather a **"Pyramid Schematic"** which provides the reader with real, sustainable, practical, and applicable solutions and methods for actually building a pyramid.

One of the reasons why the Geometry of the Pyramid was used for the first large-scale structures built by Humans is because the Pyramid Structure can seemingly be infinitely scaled up in size and proportion and still maintains its stability. Likewise, The Blueprints or **"Black-prints"** for Pyramid Building provided within this book can be scaled up in size and proportion to construct massive Pyramids like the Ancients. The following sections outline the blueprints, diagrams, and schematics for the two Pyramid building Techniques that were tested for this Project.

5.1. The Binary Arch Pyramid

The Binary Arch "Pyramid" is really more of a Dual-Arch Nubian Vault that approximates the shape and form of a Pyramid. The Binary Arch Pyramid is constructed using two Nubian Vault Arches crossing and intersecting at 90° degrees. We have affectionately named the Binary Arch Pyramid the **"PYRAMID OF THE HIGH NOON MOON"** because the Arch shape is similar to the Crescent Moon. The appearance of the Binary-Arch Pyramid can be likened to "**a wheel within a wheel upon the Earth**" spoken of in the Bible book of Ezekiel chapter 1 verse 16 which describes the chariot of "God" called a **Merkabah**, meaning "chariot" or "to ride", denoting a CRAFT. The "Merkaba" is also depicted as a double Tetrahedron Pyramid in the Metatron Cube.

The first step in the design of the Binary Arch Pyramid is to decide on the size of the base. The Binary Arch Pyramid is built on a square foundation, and the Arches cross diagonally from each corner of the square foundation. We can use the trigonometric functions discussed in the previous chapter to calculate the length and width of our square foundation based on the desired width of the Arches.

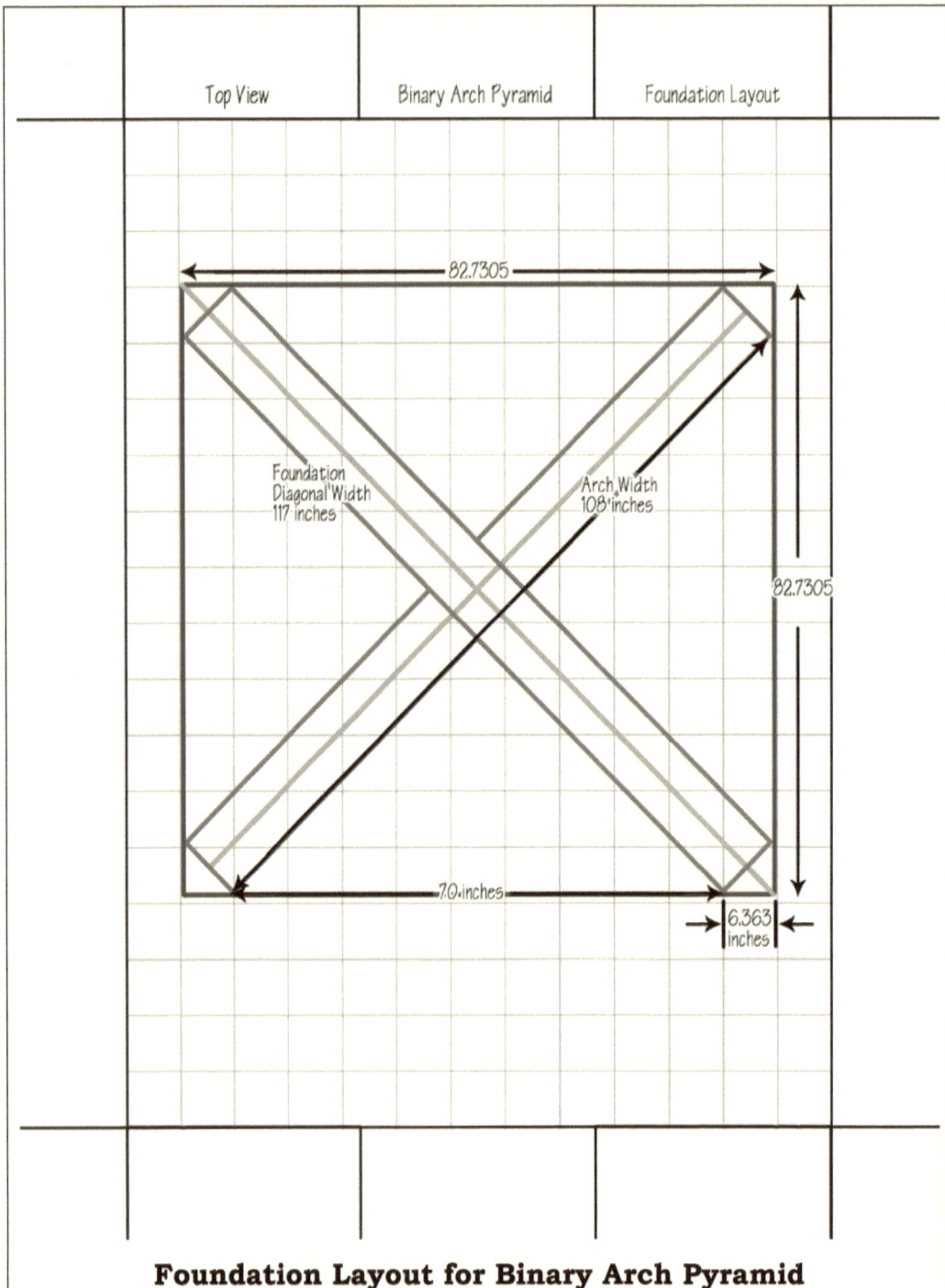

Top View | Binary Arch Pyramid | Foundation Layout

82.7305

Foundation
Diagonal Width
117 inches

Arch Width
108 inches

82.7305

7.0 inches

6.363
inches

Foundation Layout for Binary Arch Pyramid

We designed the Binary Arch Pyramid with two Arches that are each 108 inches wide (9' ft, 2.74 meters) and 75 inches in height (6.25 ft, 1.9 meters). The Arches span diagonally across a square base. Since our Arches are 108 inches wide, we can use the trigonometric functions to calculate the minimum required width and length of the square base to be 82.7 inches by 82.7 inches (6.9 ft × 6.9 ft, 2.1 meters × 2.1 meters). In order to build the Arches, we will need to build wood formwork on which the bricks can lay. The wood formwork is built by nailing together 2"×4" wood planks which are the desired height and width of the Arches. The wood formwork should also include wood planks at 15°, 30°, 45°, 60°, and 75° in order to help support the weight of the bricks. A piece of plywood the length of the total Arch is nailed across the wood form creating a "ramp" on which the bricks can be laid. We can use the mathematic tools previously described to calculate the length of the wood support arms and the length of the plywood Arch. The wood formwork is screwed onto wood plank footings to help balance the formwork. After the bricks are laid on the formwork and the Arch dries, the wood plank footings are unscrewed, the formwork is lowered and removed, and the brick Arch remains standing.

Side View	Binary Arch Pyramid	Formwork

0. Planks to raise framework by 3 inches
1. Base Plank, 108 inches long
2. 15 degree plank, 50.8 inches long
3. 30 degree plank, 50.7 inches long
4. 45 degree plank, 53.6 inches long
5. 60 degree plank, 59.9 inches long
6. 75 degree plank, 69.1 inches long
7. Vertical 90 degree plank, 75 inches long
8. Plywood Arch, 193.6 inches long

Formwork Design for Binary Arch Pyramid

| Side View | Binary Arch Pyramid | Brick Arch on Formwork |

Bricks laid across the Arch formwork of Binary Arch Pyramid

Side View	Binary Arch Pyramid	Brick Arch

Brick Arch after removing formwork of Binary Arch Pyramid

3D View	Binary Arch Pyramid	Finished Brick Arches

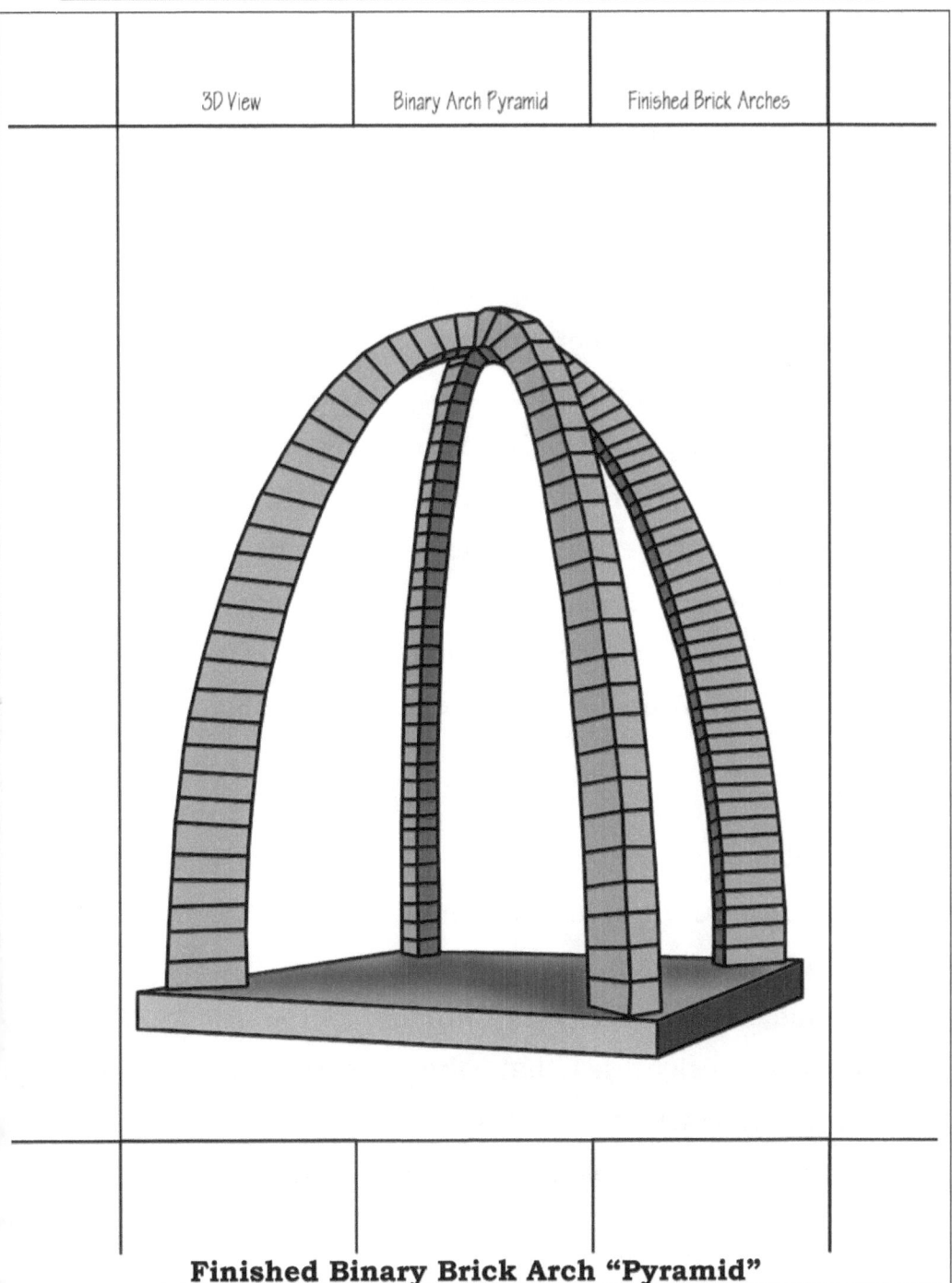

Finished Binary Brick Arch "Pyramid"

The Finished Binary Brick Arch Pyramid has two Nubian Vault Arches intersecting at a right angle running diagonally from corner-to-corner on a square base. The Finished Binary Brick Arch Pyramid approximates a Pyramidal shape. Although the design is open, walls can be created using wood, plaster, glass, or bricks. Brick walls can be built using the squinch and cloister method previously discussed. If the Binary Arch Pyramid is topped with a Pyramidal pointed capstone built at an angle and ratio equal to the height and width of the Arches, then the Finished Binary Brick Arch Pyramid will more closely resemble a Pyramidal form However, the Binary Brick Arch "Pyramid" is in fact a Nubian Domical Arch.

The advantage of the Binary Arch Pyramid design is that a tall structure which approximates a Pyramid form can be constructed quickly, inexpensively, and with relatively few materials. The disadvantage of the Binary Arch Pyramid design is that the construction requires formwork which makes scaling up the construction to larger sizes less stable and more challenging.

5.2. The Squared-Circle Stepped Pyramid

The Squared-Circle Stepped Pyramid combines the geometric forms of the Circle and the Square using the corbelling method of laying successive courses of offset bricks to construct a true Pyramid based on the Golden Triangle. The design of

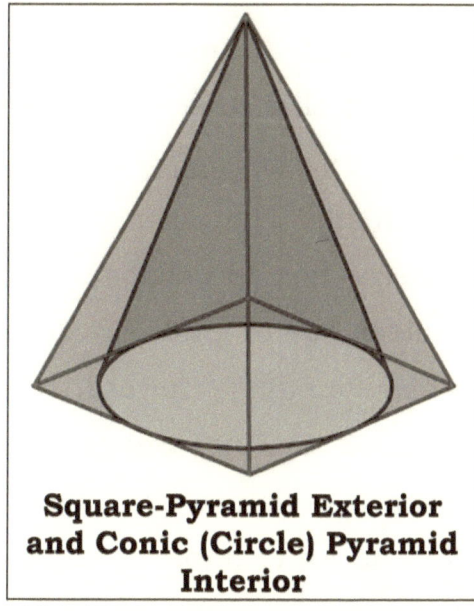

Square-Pyramid Exterior and Conic (Circle) Pyramid Interior

the Squared-Circle Stepped Pyramid calls for the exterior of the structure to be based on the "Square" forming an exterior Square Pyramid, while the interior of the structure is based on the "Circle" forming an interior Conic Pyramid. The design of the "Binary Arch" Pyramid has vertical "Arches"; however, the design of the Squared-Circle Stepped Pyramid has internal horizontal "Arches" (the circle") which support the exterior courses of offset bricks. The process of gradually decreasing the size of the square layers while building upward is perhaps the most intuitive method for constructing a pyramid shape. The "Stepping" or gradual decrease in the size of the square layers is the method that Imhotep, the Nubians, the Olmecs, and the Mesoamericans used to build stepped pyramids. Even the smooth-sided pyramids are built with the same technique, but then are finished with stones cut at angles to finish the "smooth" sides. The problem that arises when a builder gradually decreases the size of the <u>square</u> layers while building upward is that eventually the compression forces will cause the construction to collapse in the center. Conversely, if a builder was to gradually decrease the size of <u>circle</u> layers while building upward, the construction would not collapse in the center because circular and round geometric shapes are very efficient against compression forces. An example of this fact is found in the answer to the currently popular critical-thinking interview question: **"Why are manhole covers round?"** There are several reasons why manhole covers are built round, but one of the most important reasons is because a round manhole cover cannot fall through its circular opening, but there is potential that a square manhole cover may fall in the opening. The inner circle or arches make the design of the Squared-Circle Stepped Pyramid scalable to the sizes of multi-level buildings. The interior circle or conic section can even be designed as a **spiral ramp** which can be used to transport building materials as the structure grows upward.

A Conic Pyramid (Circle-base Round "Pyramid") could easily be constructed without the construction collapsing in the center by gradually decreasing the size of circle layers while building upward. Thus, the solution to constructing a Stepped-Pyramid with a square-base like the ancient pyramids of Imhotep, the Nubians, and the Olmecs would be to "Square-the-Circle" – both literally and figuratively. The "Squared-Circle" Stepped Pyramid combines the strength and durability of constructing using the Circle, and also gives the visible appearance of an ancient Pyramid using the Square.

The design of the "Squared-Circle" Stepped Pyramid is a **"union of the two lands"** of the circle and the square. The term **"squaring the circle"** was associated with the Ancient **Alchemical great work (Magnum Opus)** of creating the **Philosopher's Stone** used to transmute base metals into gold and obtain immortality. Since most of the language associated with "The

Alchemical Symbol for "Squaring the Circle"

Philosopher's stone" and "Squaring the Circle" in Alchemy is symbolic and allegorical, it is no surprise that this method has actual practical applications in the design and construction of Pyramids.

We have affectionately named the Squared-Circle Stepped Pyramid the **"PYRAMID OF THE MIDNIGHT SUN"** because the capstone of the Pyramid is topped with a light that acts as a "sun" at night. The test design for the Squared-Circle Stepped Pyramid that was constructed for this project is for a structure of base 50 inches (4.17ft or 1.27 meters), height of 75 inches (6.25ft or 1.91 meters) with 19 courses of bricks and a Plexiglas capstone. However, if we choose a base of 32 ft. and a height of 48 ft., the Squared-Circle Stepped Pyramid could be scaled up to design a 4-story 1500 Sq ft home. The interior of the structure will be a **Conic Pyramid** called a **BURUTU** in the **Twi** language of West Africa. If we set each floor to have a height of 12 ft., then we can design the 1st floor to have a diameter of 32 feet with 804 square feet of space; the 2nd floor to have a diameter of 24 feet with 452 square feet of space; the 3rd floor to have a diameter of 16 feet with 212 square feet of space; and the 4th floor to have a diameter of 8 feet with 100 square feet of space.

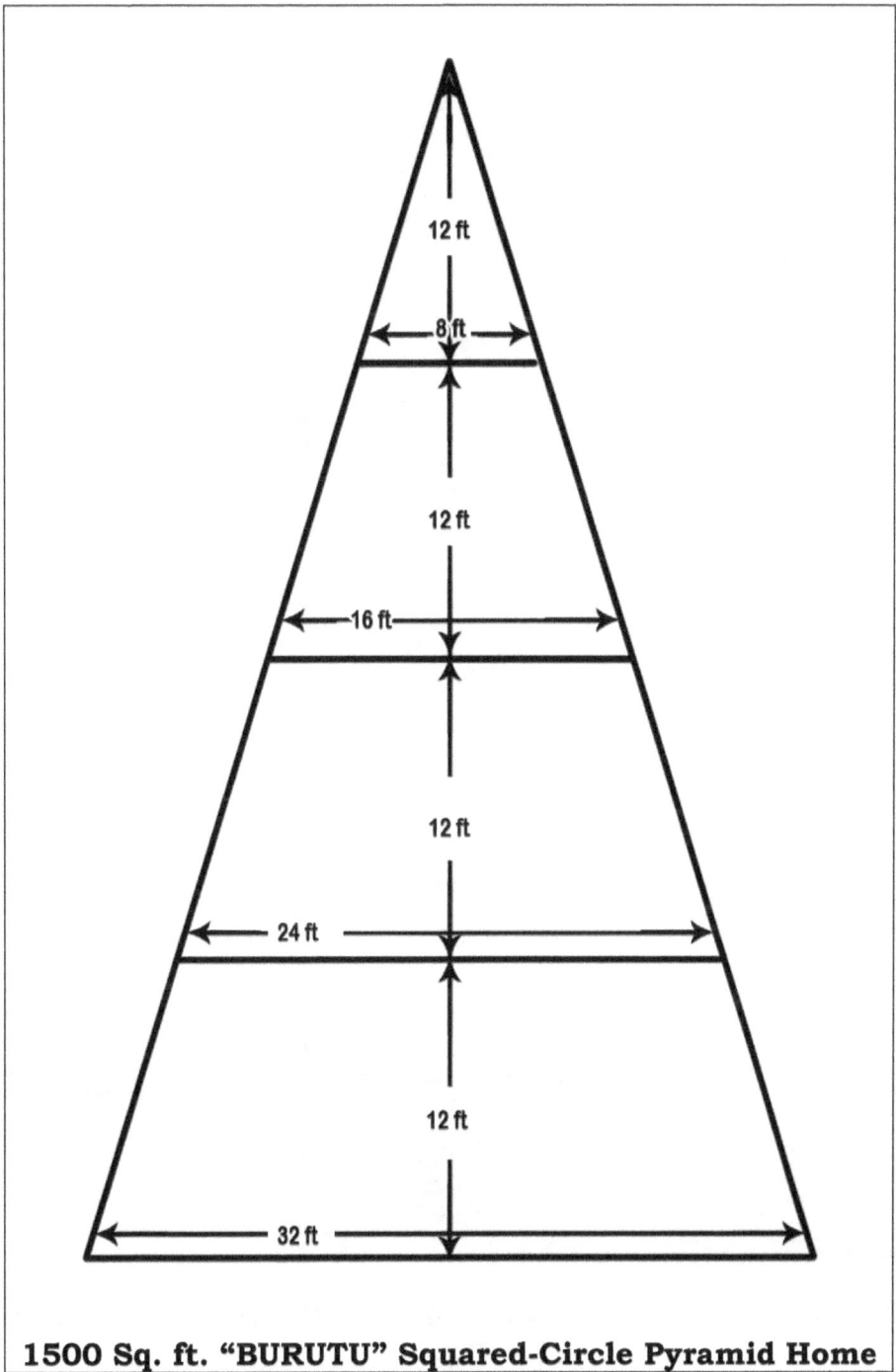

1500 Sq. ft. "BURUTU" Squared-Circle Pyramid Home

2ND FLOOR

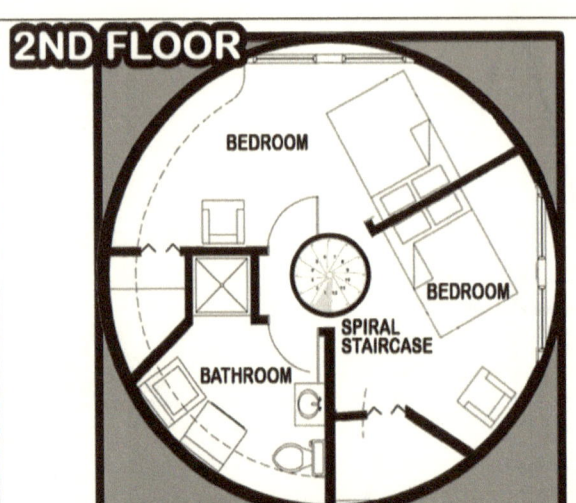

4TH FLOOR

1ST FLOOR

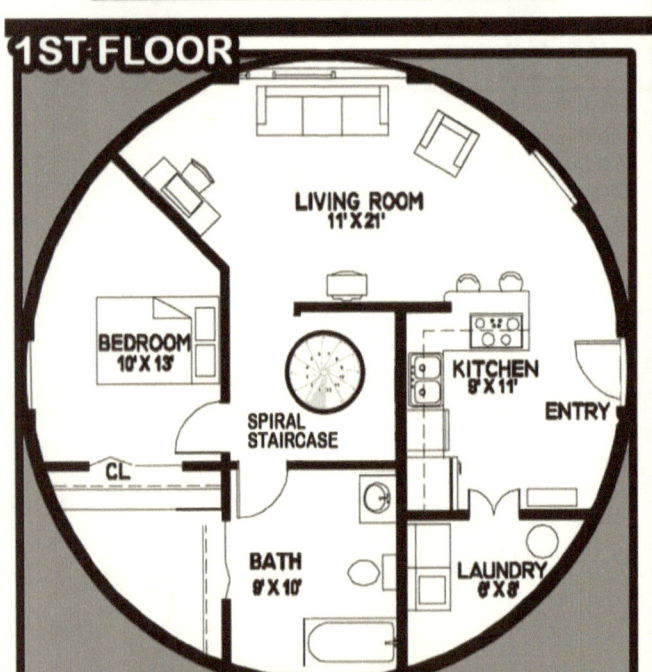

3RD FLOOR

Sample floor plans for 1500 Sq. ft. BURUTU Squared-Circle Pyramid Home

	3D View	Squared-Circle Pyramid	Finished Offset Brick Courses	

Squared-Circle Stepped Pyramid
"Pyramid of the Midnight Sun" Schematics

| Top View | Squared-Circle Step Pyramid | LAYER 1 |

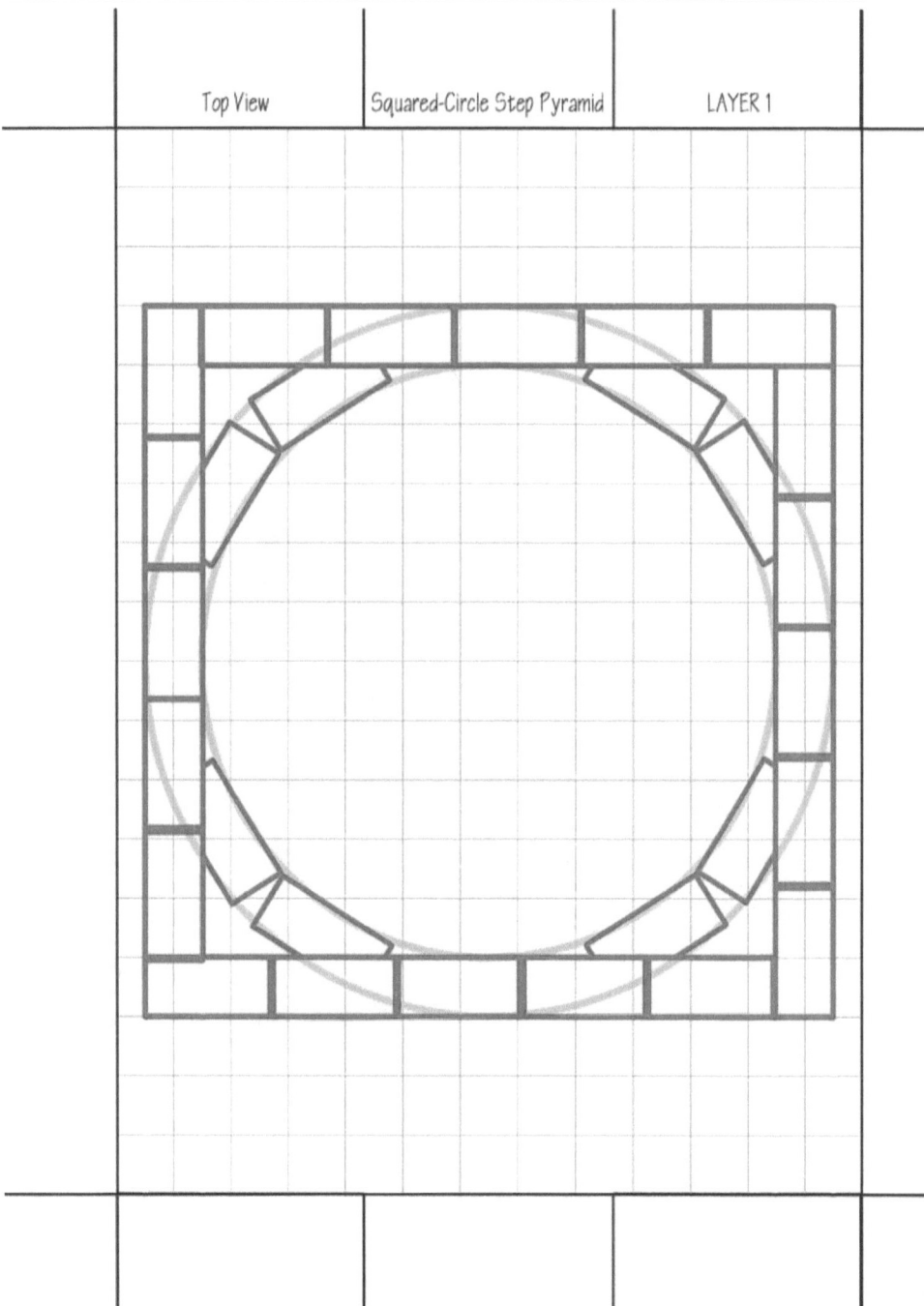

| Top View | Squared-Circle Step Pyramid | LAYER 2 |

| Top View | Squared-Circle Step Pyramid | LAYER 3 |

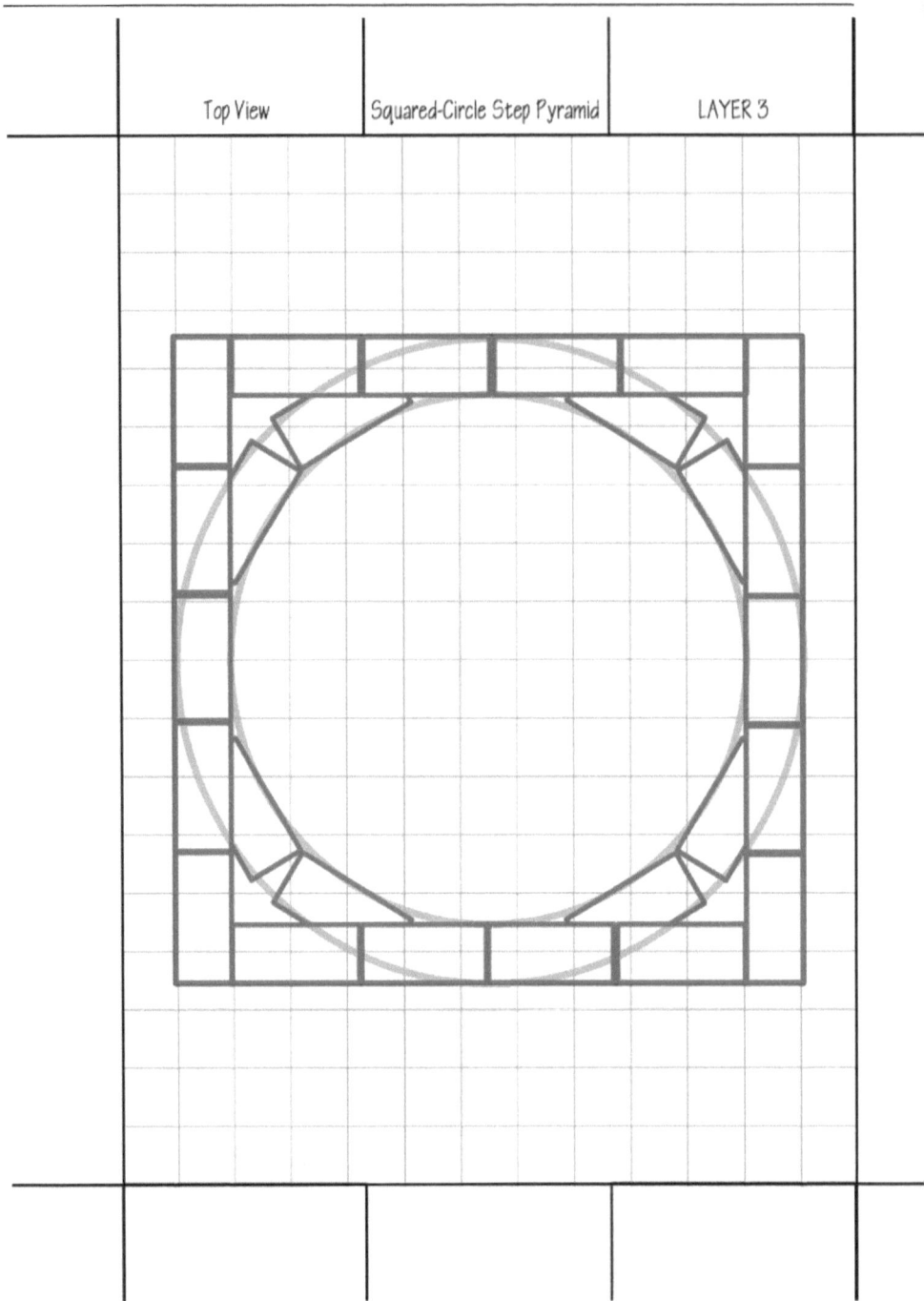

| Top View | Squared-Circle Step Pyramid | LAYER 4 |

Top View	Squared-Circle Step Pyramid	LAYER 5

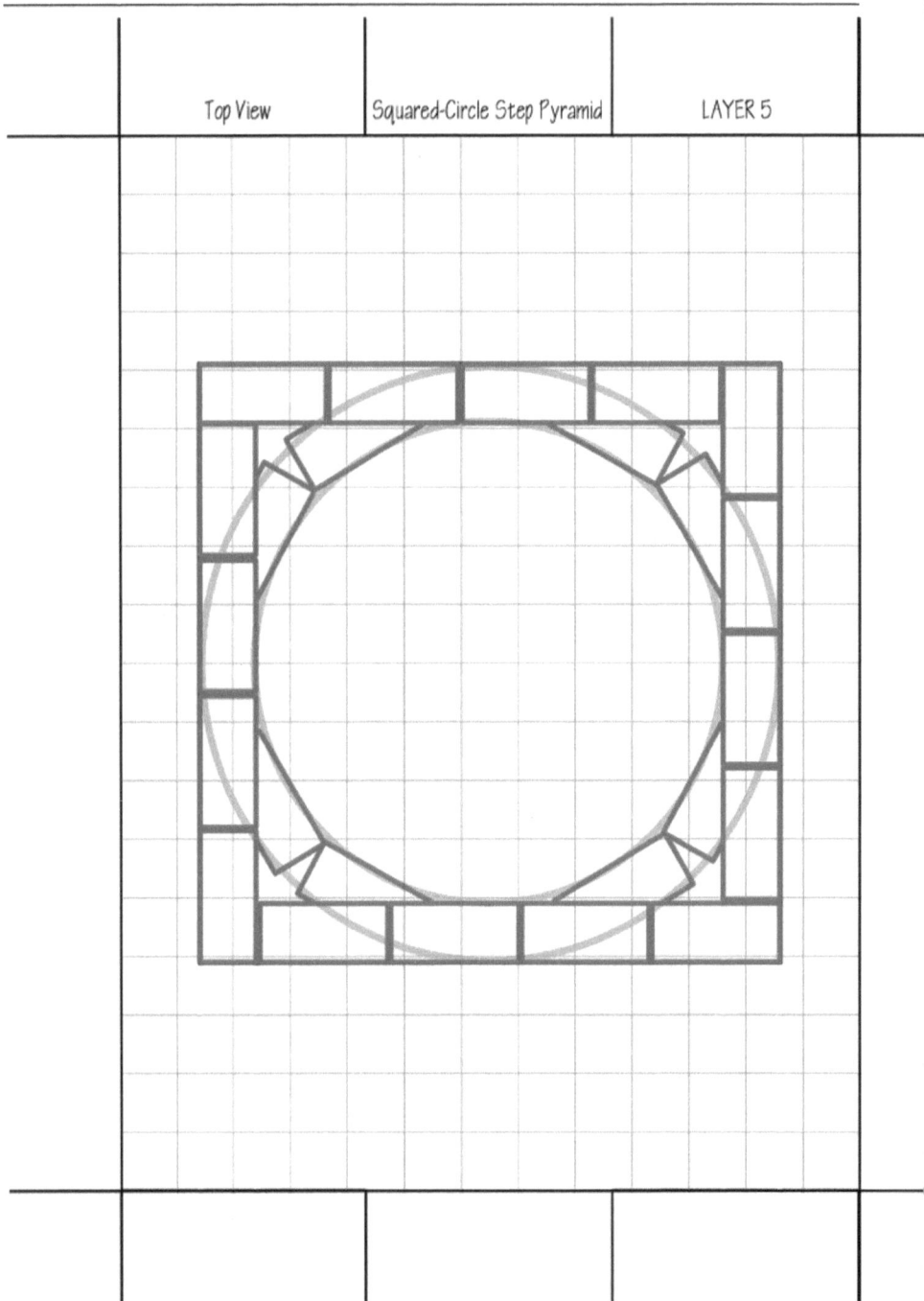

| Top View | Squared-Circle Step Pyramid | LAYER 6 |

| Top View | Squared-Circle Step Pyramid | LAYER 7 |

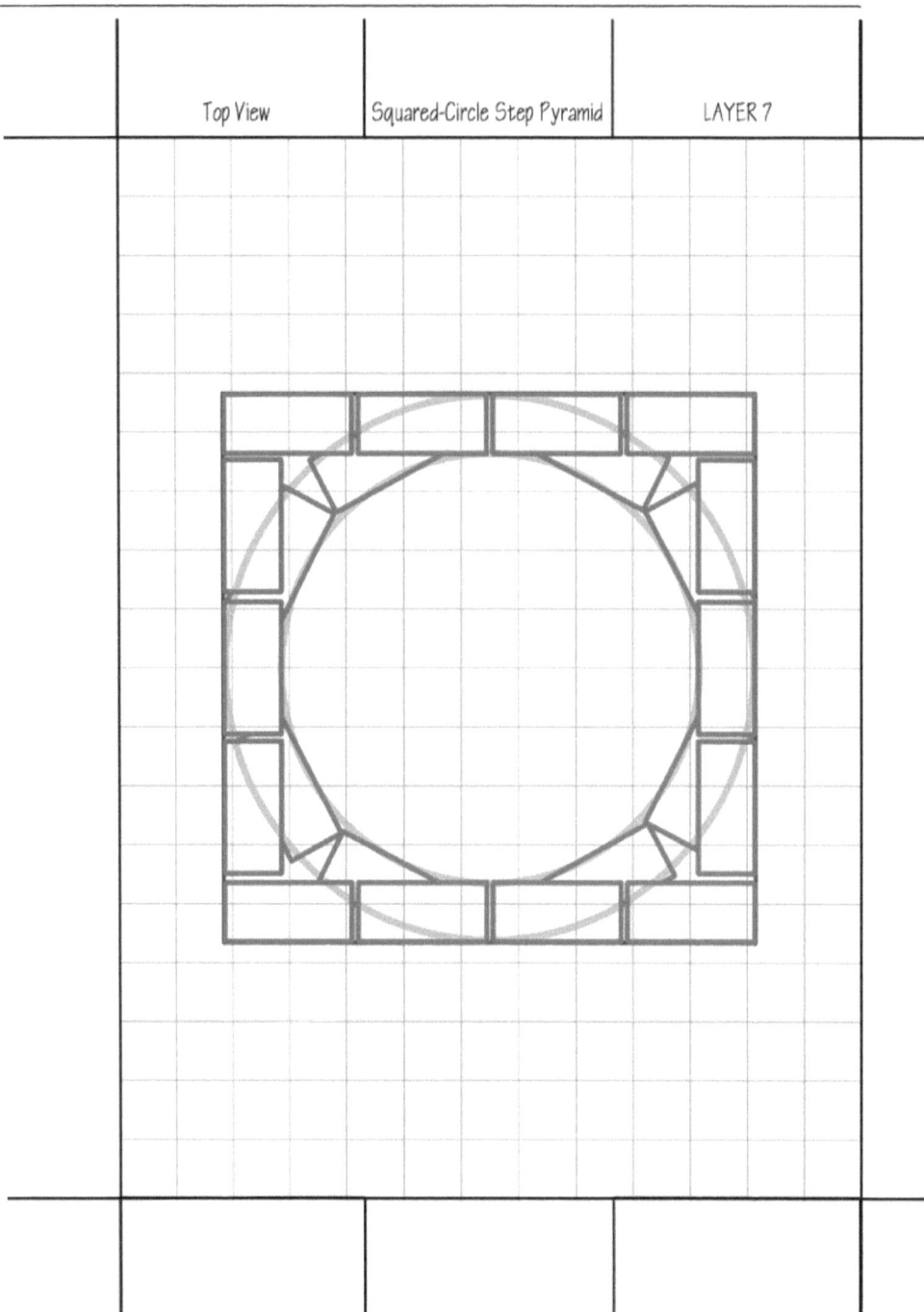

| Top View | Squared-Circle Step Pyramid | LAYER 8 |

| Top View | Squared-Circle Step Pyramid | LAYER 9 |

| Top View | Squared-Circle Step Pyramid | LAYER 10 |

| Top View | Squared-Circle Step Pyramid | LAYER 11 |

| Top View | Squared-Circle Step Pyramid | LAYER 12 |

Top View	Squared-Circle Step Pyramid	LAYER 13

| Top View | Squared-Circle Step Pyramid | LAYER 14 |

| Top View | Squared-Circle Step Pyramid | LAYER 15 |

| Top View | Squared-Circle Step Pyramid | LAYER 16 |

Top View	Squared-Circle Step Pyramid	LAYER 17

| Top View | Squared-Circle Step Pyramid | LAYER 18 |

| Top View | Squared-Circle Step Pyramid | LAYER 19 |

The capstone of a Pyramid is basically a miniature version of the Pyramid built at the same proportions. The design of the capstone for the Squared-Circle stepped Pyramid is done by calculating the width of the final layer of bricks, then using the Trigonometric functions to calculate the height of the capstone based on the offset distance of the corbelled brick layers that was maintained during construction. Capstones are usually one solid form, and thus can be constructed by creating a mold to cast the capstone based on the Pyramid's proportions. Polygons with more than 5 sides can be used to create forms to mold capstones of different angles. If the capstone will be fixed directly to the last layer of bricks, then the base of the capstone will be equal to the width of the last layer of bricks. However, for this project, we designed the capstone to give the appearance that it is elevated or floating above the Pyramid. To design the capstone to appear elevated above the Pyramid, pick an elevation gap distance that is less than the total height of the Pyramid minus the total height of the brick layers. Using the chosen elevation gap distance, the Trigonometric functions can be used to calculate the base and height of the capstone. For this project, the capstone was designed based on the Golden Triangle with a base of 6 inches to be built from Plexiglas so that a light could shine out of the Capstone Pyramidion representing the concepts associated with the ancient BenBen Stone on top of the Primordial Mound. The following table lists the dimensions used for the Squared-Circle Pyramid constructed for this project.

	LAYER	WIDTH (inches)	HEIGHT (inches)
	\multicolumn{3}{c}{**SQUARED-CIRCLE PYRAMID DIMENSIONS**}		

	LAYER	WIDTH (inches)	HEIGHT (inches)
BRICKS	1	50	3
	2	48	6
	3	46	9
	4	44	12
	5	42	15
	6	40	18
	7	38	21
	8	36	24
	9	34	27
	10	32	30
	11	30	33
	12	28	36
	13	26	39
	14	24	42
	15	22	45
	16	20	48
	17	18	51
	18	16	54
	19	14	57
GAP		12	60
		10	63
		8	66
CAPSTONE		6	69
		4	72
		2	75

POLYGONS CAN BE USED FOR CAPSTONE MOLDS

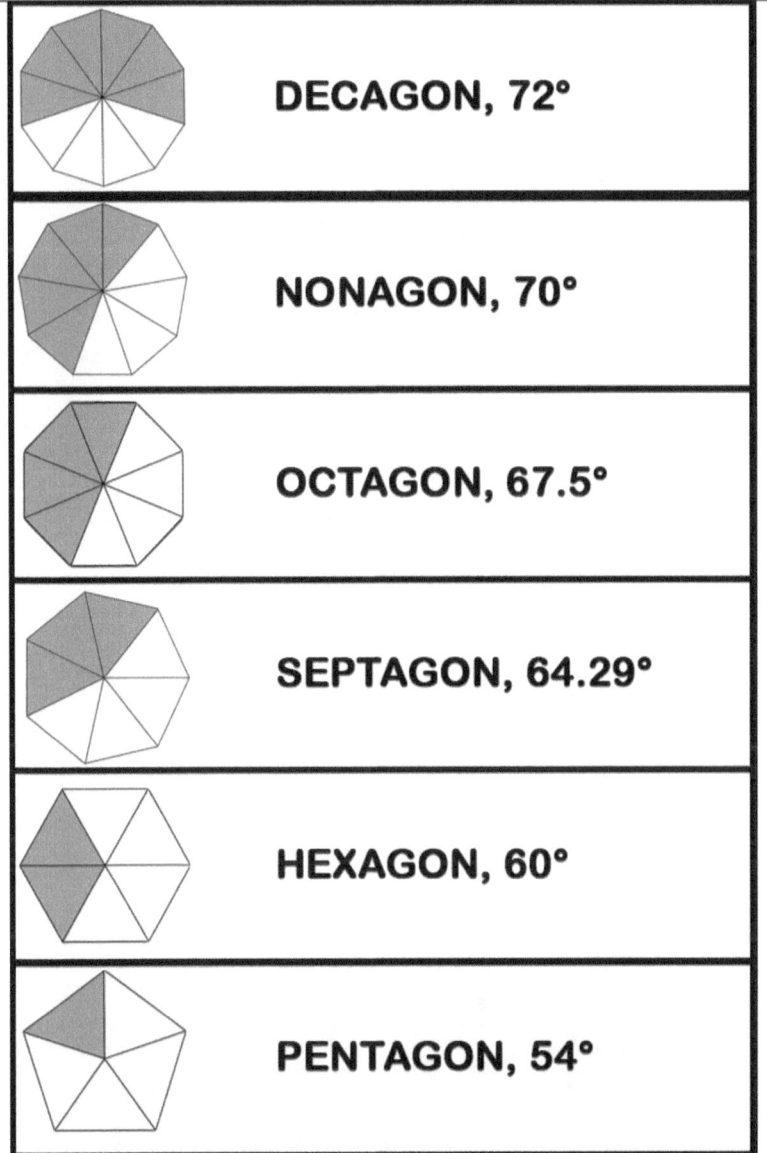

	DECAGON, 72°
	NONAGON, 70°
	OCTAGON, 67.5°
	SEPTAGON, 64.29°
	HEXAGON, 60°
	PENTAGON, 54°

If a polygon is designed with all of its sides equal to the base of the capstone, then 4 triangles of the Polygon can be used as a mold and/or form for Capstones to fit Pyramids with different angles.

Above: Squared-Circle Step Pyramid with 19 courses of bricks and lit elevated Capstone light

OSIADAN

6.0. CONSTRUCTION INSTRUCTIONS

Now that we have developed our schematics for our two prototype Pyramids, we can begin the construction process. In this chapter we will provide step-by-step instructions to construct our "Pyramid of the Midnight Sun" (the squared-circle stepped pyramid) and the "Pyramid of the High-Noon Moon" (the dual-arch Pyramid).

Building materials such as Bricks and Mortar are relatively inexpensive to obtain, but these materials can also be easily made by using the raw material of the Earth. In addition to providing the instructions to construct the Pyramids described in chapter 5, we also provide "recipes" to make Bricks and Mortar in this chapter.

Always remember to keep safety first, and check the codes and laws of the area in which you are building to ensure that your finished structure will be in accordance with all building and safety codes.

So let's get to work!

6.1. Tools of the Craft

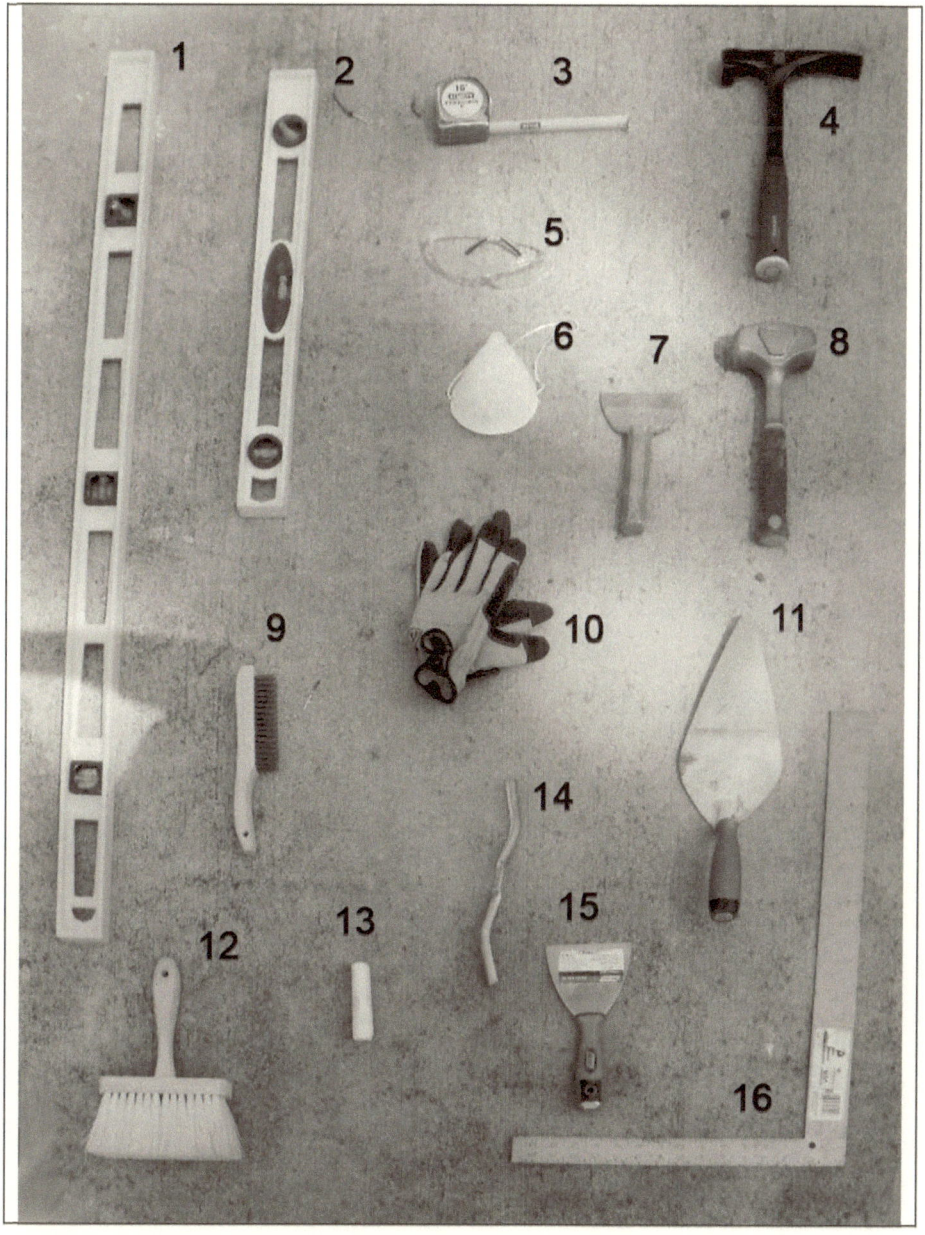

Tools of the Craft (continued)

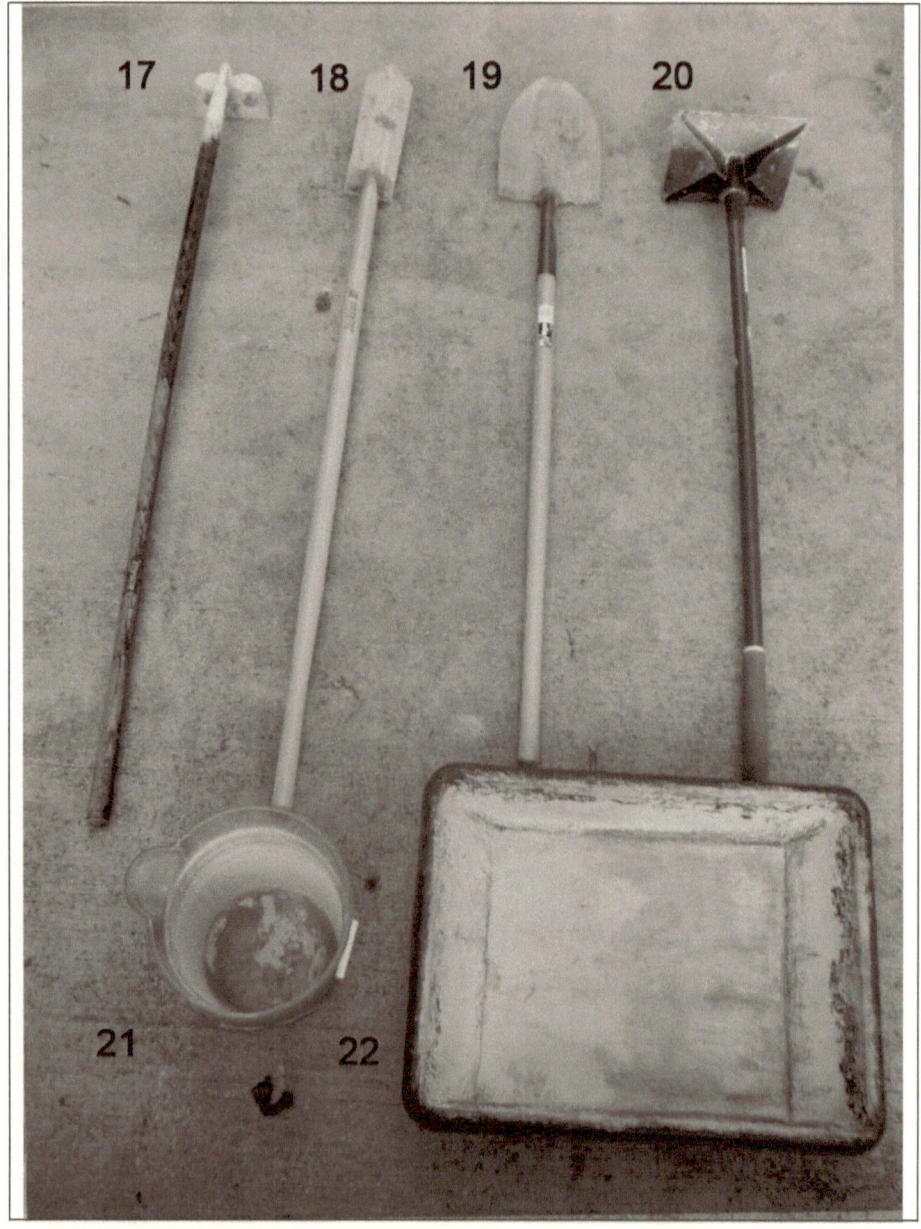

Having the proper tools is of the utmost importance in order to facilitate construction. The following is the list of tools used for this project as displayed on the previous pages:

01. 4 ft Level
02. 2 ft Level
03. 16 ft Measuring Tape
04. Masonry Brick hammer
05. Protective Eye wear
06. Protective Face Mask
07. Brick Set Chisel
08. Small Sledge Hammer
09. Steel Wire Brush
10. Work Gloves
11. Masonry Mortar Trowel
12. Mason's Brush
13. Chalk
14. Mortar Joint Tool
15. Scraper
16. 90° Framing Square
17. Mortar Hoe
18. Long Handle Spade Shovel
19. Round Mouth Shovel
20. Hand Tamper
21. 8 Gallon Bucket
22. Mortar Mixing Box

6.2. How to Make Bricks

INGRIDIENTS NEEDED:

* Soil   * Water   * Sun Light/Fire

* Sand   * Clay   * Wood Planks

TOOLS NEEDED:

* Shovel   * Hammer   * Nails

* Saw   * Ruler   * Glass Cup

* Teaspoon   * Mixing Container

STEP 1: TEST THE SOIL

- Mix 10 teaspoons of soil in 1 cup of water.

- Stir the mixture and leave it to sit overnight.

- The next day you will see one layer of clay on the top and a layer of sand on the bottom. Use a ruler to measure the height of the layer of clay and the height of the layer of sand.

- The ratio of the height of the Sand layer to the height of the Clay layer should be at most "**70 percent sand to 30 percent clay**" and no less than "50 percent sand to 50 percent clay"

- If the ratio of sand to clay is appropriate, then the soil is good for brick making. If the ratio is not sufficient, then add clay or sand to the soil as needed.

STEP 2: MAKE A BRICK MOLD FOR THE MIXTURE

- Brick molds can be constructed in various sizes using wood and nails. The brick mold can be constructed for one brick or multiple bricks.

- Use your saw, wood planks, hammer, and nails to build your mold. The dimensions of a mold for an average size brick would be 8 inches long (203 mm), 4 inches wide (102 mm), and 2.25 inches high (57 mm). The dimensions of a mold for a traditional adobe brick would be 14 inches long by 10 inches wide by 4 inches high

STEP 3: PREPARE THE BRICK MIXTURE

- The Brick Mixture can be mixed in a wheel barrow or any container large enough. You can also dig a large pit to mix the Brick Mixture.

- Put the soil into the container to be mixed, add water to the soil, and thoroughly mix the mixture so that it has a muddy consistency that is not too watery and still able to hold its shape.

STEP 4: FORM THE BRICK

- Pour the mixture into the brick mold and be sure to press the mixture down to fill all of the corners.
- Smooth off the edges with your hand or a shovel, and let the mixture set in the mold for 30 minutes.
- After 30 minutes of setting, the brick should be able to slide out of the mold as a raw brick

STEP 5: SUN DRY THE BRICK

- Cover the area where the brick will dry with sand or straw to keep the wet bricks from binding to the surface of the drying area.
- Leave the raw brick to dry in the Sun for 3 weeks rotating the brick onto a different edge every 3 days.
- After 3 weeks all of the exposed edges should turn white and the brick should be ready for use

STEP 6: FIRE DRY THE BRICK

- To make a fired red clay brick, cure the brick in a kiln oven at 1800 degrees Fahrenheit for 7 additional days.

ARI-KAT: IMHOTEP'S METHOD FOR FABRICATING LIMESTONE BLOCKS

As we learned earlier, after visiting the **Temple of Khnum**, the **Architect Imhotep** was given a formula to fabricate Limestone blocks using a method that is similar to the method we use to make modern bricks. However, Imhotep's fabricated Limestone blocks were much more durable then our modern bricks. The steps below are a general outline to how Imhotep may have fabricated a Limestone Block.

Ingredients:

* Soft Limestone   * Natron Salt

* Lime   * Water

STEP 1:

Disaggregate the Soft Limestone by diluting it in water until the lime and clay separate into a layer of mud on the top and a layer of fossil shells on the bottom.

STEP 2:

Add Natron salt (sodium carbonate) to the disaggregated mixture. Natron Salt is used as a hardening agent and was also used in the mummification process

STEP 3:

Add Lime to the mixture to serve as a binding agent. Lime can be created by burning limestone, dolomite, plants, and other sedimentary rocks to ashes and adding it to the mixture.

STEP 4:

Create Caustic Soda by mixing Lime, Natron Salt, and water in the mixture. The Caustic Soda serves as the catalyst for the chemical reaction to transform the other materials.

STEP 5:

Thoroughly mix the complete mixture in a container until it becomes a paste.

STEP 6:

Add limestone rubble, silt, and fossil shells to the mixture to create a concrete paste.

STEP 7:

Cover the block mold with oil, and then pour and pack the Limestone concrete paste into the mold. Let the block dry for several weeks.

6.3. How to Make Mortar

INGRIDIENTS NEEDED:

* Portland Cement   * Sand

* Dry Builder's Lime   * Water

TOOLS NEEDED:

* Trowel   * Mortar hoe   * Mixing Container

STEP 1:

Combine two parts Portland cement with one part dry builder's lime in a container

STEP 2:

Mix two to three parts sand to the dry Portland cement and lime mixture from step 1.

STEP 3:

Mix in water with the dry ingredients until it reaches a muddy consistency to where it easily slides off the trowel.

6.4. Laying the Foundation

The Ancient Pyramid Builders learned the importance of building on a solid rock foundation after the construction of Huni's Pyramid at Meidum. The Smooth sides of Huni's Pyramid at Meidum was built on the sand as opposed to on solid rock or gravel which is one of the factors that led to the smooth sides of the Pyramid at Meidum to collapsing. Preparing the construction site and laying the Foundation sets the tone for the entire building project. Builders can spend just as much time preparing a site for construction and laying a stable, solid, level foundation as they spend actually building on the foundation. The following points and steps should be considered when preparing a construction site and laying a foundation.

STEP 0:

Check the laws and codes of your area to find out the necessary thickness and depth of the foundation and footings based on the size of your building project. Adhering to Building codes ensures the safety and stability of the finished construction.

STEP 1:

Survey the Land and notice the natural features. Try to insure that your building project works in conjunction with the natural features of the area.

STEP 2:

Mark the project area based on your designs using rope, string, or paint. Remove all unnecessary debris, plants, and other vegetation or structures to clear and clean the project building area.

STEP 3:

Begin digging and excavating the building area to lay a sub-base. Depending on the size of your building project, it may be necessary to also include "footings". Use a Level and Measure the slope of the land to determine if Grading is necessary. Grading is the process of leveling sloped land. If Grading is necessary, the excavation process can be used to level the land.

STEP 4:

Add a layer of compactable gravel 6 inches thick and use a Tamper tool to compress it to 4 inches thick.

STEP 5:

Place wood planks in the excavated area on top of the compact gravel sub-base to the size and shape of the foundation. Since wood planks are used, rectangles, squares, and triangles are the easiest shapes to make for a foundation; circular foundations are more complicated.

STEP 6:

Mix concrete the concrete for the foundation slab. For smaller projects, it is possible to mix the concrete using a Mortar Hoe in a Mortar Box. However, larger projects will require the concrete to be mixed by a machine so that it does not prematurely dry out. Concrete that is mixed with too much or too little water will be weak. If you can squeeze the mixed concrete in your hand and it retains its shape, then the concrete has been mixed to the correct consistency.

STEP 7:

Pour the concrete into the wood form and use another wood board to smooth the wet concrete. Allow the concrete slab to set and dry for a couple of days before building.

6.5. Building the Binary Arch Pyramid

In Masonry, building Brick Arches that are strong and sturdy requires highly accurate mathematics; precise craftsmanship; and is considered a very advanced and specialized skill. Building two Brick Arches intersecting at a 90° degree angle transcends one of the most advanced concepts in Masonry, but as Pyramid Builders, this is the point where we initiate our building process.

STEP 0:

Pour a concrete slab foundation for the building project. Use chalk, paint, or rope to mark the "Foundation Layout" (page 132) of the Binary Arch Pyramid onto the foundation which it will be built.

STEP 1:

Use a hammer and nails to construct the wood Formwork for Binary Arch Pyramid (page 134). Do not attach the bottom support planks (Labeled 0 on page 134) at this time.

STEP 2:

Place the wood formwork diagonally onto the marked area of your foundation. Screw the wood formwork to the bottom support planks (Labeled 0 on page 134) at this time. At this point the wood formwork should be standing on the bottom support planks and position diagonally from corner-to-corner on the foundation ready for the bricks arch to be built.

STEP 3:

Mix Mortar using your Mortar Hoe and Mortar box

STEP 4:

Using your Masonry Mortar Trowel, place a bed of mortar at each end of the wood formwork. Apply mortar to both faces of a brick and lay the cornerstone of one Arch at one end of the wood framework ensuring that the edge of the brick is square with the framework. Use your level to make sure the brick is level, and tap the brick using the Trowel to level the brick if and where needed. Repeat this process to lay the Brick Cornerstone of the Arch at the other end of the formwork.

STEP 5:

Use your Trowel to place a layer of mortar on top of the cornerstone brick that was laid. Apply mortar to both faces of another Brick, and place it on top of the previous brick using your level to ensure that all the bricks are in line and square with each other and the formwork. The bricks should be placed so that their face is laying on the formwork, and it may be necessary to place a small pebble in between the bricks to create the wedging action needed to achieve this effect. Repeat this process to continue laying bricks against the wood formwork alternating sides every 10 bricks until the brick Arch is built across the wood formwork Arch as shown on page 135.

STEP 6:

Allow the brick Arch to set and dry for two weeks. After the Brick Arch has set and dried, remove the wood formwork by unscrewing the wood formwork from the bottom support planks. Slide the bottom support planks from underneath the wood formwork and the wood formwork can be lowered and removed from underneath the Arch. It may be possible that some mortar dried on the wood formwork which will mean you will have to use your brick hammer and a small chisel to chip the mortar off of the formwork before it can be lowered and removed from underneath the Arch. After the formwork is removed, the Brick Arch should still be standing as shown on page 136.

STEP 7:

Rotate the wood formwork 90° degrees and place the wood formwork diagonally into the other marked area of your foundation to build the 2nd Arch. Screw the wood formwork to the bottom support planks so that now the wood formwork is standing on the bottom support planks and position diagonally from corner-to-corner and intersecting at 90° degrees with the brick Arch that was just built.

STEP 8:

Repeat step 3, step 4, step 5, and step 6 to build the second Brick Arch across the wood formwork and remove the wood formwork after the second Brick Arch has set and dried for two weeks to obtain the Binary Arch Pyramid as shown on page 137.

STEP 9:

(Optional) Use your Trowel to cover the Binary Arch Pyramid with Hard coat stucco to help reinforce, strengthen, and water-proof the structure.

In Ancient times, after the Mason would complete a Brick Arch project, he was required to **"stand under the Arch"** (**under-stand**) to show that his work was sturdy and safe. When the Nubians would build their Nubian Vault Arches, they would stand under the Arch and **"stand on top of the Arch"** (**over-stand**) to show that the Nubian Vault Arch was sturdy and safe enough to support the weight of the Mason.   We must reemphasize the importance of safety, and thus we do not suggest or encourage anyone to stand under or stand on an Arch that they built.   However, the strength of the Arch can be tested without putting yourself in harm's way by standing on a ladder and placing bricks or weights on the top of the Arch greater than or equal to your own weight to test the strength of the Arch.

The four sides of the Binary Arch Pyramid can be closed off by using successively smaller Arches creating a Cloister or "Domical Vault".   The Binary Arch Pyramid can also be topped with a capstone which will create an end result which has an appearance similar to the "Bent Pyramid" of Sneferu.

6.6. Building the Squared-Circle Pyramid

The Masonry techniques required to build the "Squared-Circle" Pyramid are the same basic techniques required to build a brick wall. Any additional resources that can be acquired on the topic of Brick Masonry would be good references in addition to the information presented here. Corbelling the successive brick layers and building the inner circle of the Pyramid/Cone structure are the only other additional Masonry skills required to successfully build the Squared-Circle Pyramid.

STEP 0:

> Pour a concrete slab foundation for the building project. Use chalk to trace out the outline of the first layer of bricks onto the concrete foundation slab.

STEP 1:

> Extend a vertical rod or pole from the center of the future Pyramid structure to the desired height. Stretch a tight line of cord or rope from the top of the rod or pole to the corners of the base of the chalk line plus the corbel offset distance. These

tight ropes will serve as guide lines to ensure that all of the corners of the bricks are in line as you build vertically upward.

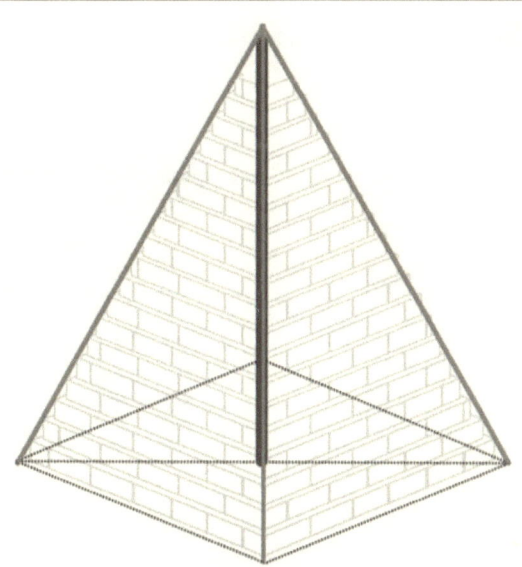

Above: Diagram showing the vertical pole extending to the height of the future structure with tight ropes leading to the corners of the base which serve as guide lines during the building process

STEP 2:

Place vertical rods at each corner and stretch a tight line of cord or rope around the rods making a square around the area. If you are using standard bricks, this square line should be 3 inches high for the first layer and you will move it up 3 inches for every layer. The nominal size of a standard brick includes the height of the mortar joints. Mortar joints are estimated to be about ⅜", so the actual size of the brick is ⅜" less than the nominal dimensions. Use a line level to make sure the line is level all the way around. This line will serve as a guide line as you build horizontally.

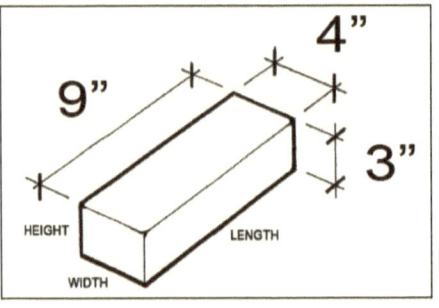

STANDARD BRICK DIMENSIONS		
	ACTUAL (inches)	NOMINAL (inches)
LENGTH	8 ⅝	9
WIDTH	3 ⅝	4
HEIGHT	2 ⅝	3

STEP 3:

Dampen the concrete foundation slab and the brick. Place a bed of mortar inside the chalk lines marked on the concrete foundation slab where the first layer of bricks will be laid. Butter one end of a Brick and place the first cornerstone brick of the first layer at one of the corners of the outer square for the first layer of the Pyramid so that the top corner of the brick touches the diagonal cord leading to the top of the Pyramid. Tap the brick down until it is level with the horizontal cord, and then use your level to check that the brick is level.

STEP 4:

Butter both ends of the next brick, and place it next to the previous brick. Tap the brick down until it is level with the horizontal line and use your level to check that the brick is level with the previous bricks laid. Continue this process until the entire square outer layer is complete. There will be some instances when it will be necessary to cut the bricks to smaller sizes to complete the building of the outer square layers. It will also be necessary to cut the bricks at angles to build the inner circle for each layer. To cut a brick straight, use your "Brick Set Chisel" and "Small Sledge Hammer" to score all four sides of a brick to the desired cut size. After scoring all four sides of the brick, firmly strike the "Brick Set Chisel" with the "Small Sledge Hammer" one last time and the brick will split at the desired cut size. To cut a brick at an angle, mark the Angled cutting line on the brick with chalk or a pencil on both sides. Score the marked cutting line on both sides of the brick using your "Brick Set Chisel". After scoring both sides of the brick, firmly strike the "Brick Set Chisel" with the "Small Sledge Hammer" one last time and the brick will split at the desired angle and cut size. Use your brick

Hammer to "square off" any ruff sides of a cut brick. There are also specialty "Brick Splitters" and "Brick Saws" that make the process of cutting bricks and blocks much easier.

STEP 5:

Tie a rope or cord to the pole or rod extending from the center of the structure. This cord will act as a "compass" to determine the inner edges of the inner circle of the Squared-Circle Pyramid layers based on the inner edge of the square layer that was laid. The brick circle will be supportive Arches at the corners of the square and the area of the brick circle will decrease as you get to the middle of the square. Use the cord/rope compass as a guide line and lay the bricks for the inner circle cutting the bricks when needed. Fill the space in the corners between the brick square and the brick circle with concrete.

STEP 6:

After the first brick-square outer-layer, and brick-circle inner-layer has been completed, you are now ready to proceed to the next layer. Measure the corbelling distance from the outer edge of the square layer and mark it using chalk or paint. For example, if the corbelling distance for each layer is 1 inch, then measure and mark 1 inch from the outer edge of the brick-square layer all the way around resulting in a chalk square drawn on the previous layer that is the corbelling distance smaller than the previous layer. This chalk line will serve as a guide line when you lay the bricks for the next layer. You can also use the horizontal cord line to help mark the corbelling distance for the next layer. As you are laying your mortar for each successive layer, make sure that it completely fills and covers the cross joints of the previous layer.

STEP 7:

Repeat steps 3 through 6 as you build up to the apex of the Squared-Circle stepped Pyramid. As you complete each layer, use your Mortar Joint Tool, Scrapper, and Mason's Brush to finish and clean up the Mortar joints of each layer. As you build up, continuously check and re-check the alignments and level of the bricks that are being laid to ensure that the finished Pyramid has a sharp precise appearance.

6.7. Placing the Capstone

The "Capstone" or Pyramidion is the apex and final piece of a Pyramid. In the Ancient Egyptian Language **Medu Neter**, the Pyramidion or Capstone was called the sacred "**BenBen**" stone. The Egyptians would cover the BenBen stone with Gold so that it would reflect the Sun's rays and appear as a Light or Sun on top of the Primordial Mound.

Building a Capstone basically involves building a mold or formwork based on the proportions of the Pyramid. Various materials can be used to make the mold including metal, wood, plastic, and glass. If making a Cement or Brick Capstone, the Pyramidion mold can be made out of wood, and the Cement or Brick mixture can be poured into the mold to form a Pyramid shaped brick. Mortar can be laid on the top layer of the Pyramid, and the Pyramidion capstone brick can then be fixed on top of the Pyramid.

For this project, the Pyramidion capstone was made out of Plexiglas so that a Light bulb could be placed inside of it to give the appearance of a light or "Sun" at night. The following are instructions for making the Plexiglas Pyramidion capstone.

INGRIDIENTS NEEDED:

* Plexiglas

* Paint

* Silicone Sealant

* Plastic Epoxy

* Painter's Tape

* Water Proofing Tape

TOOLS NEEDED:

* Drill   * Plexiglas Cutting Knife   * Sharpie Marker

* Ruler   * Compass

STEP 1:

Use a sharpie marker and a ruler to draw four triangles onto the Plexiglas. Our "Squared-Circle" Pyramid is based on the Golden Triangle, so our 4 Triangles have a base of 6 inches, a height of 9 inches. The Decagon on page 167 with sides of 6 inches can be used to help draw the four triangles on the Plexiglas. Also, draw one square on the Plexiglas with sides of 6 inches. Use your compass to draw a circle in the middle of the square with a diameter of 3 inches. The square will serve as the base of the Pyramidion Capstone and the circle will be cut out to allow a light bulb to fit inside the capstone.

STEP 2:

Cutting along the lines you drew in step 1, use your Plexiglas cutting knife and a ruler to cut out the four Triangles and square.

STEP 3:

Use the Plastic Epoxy to glue the four triangles together onto the base square to form a Pyramidion shape.

STEP 3:

Use the Silicone sealant and waterproofing tape to waterproof and seal the interior corners and edges of the Plexiglas Pyramidion.

STEP 4:

Use the painter's tape to tape the middle section of each side face of the Plexiglas Pyramidion. Add 2 to 3 coats of Paint to the corner sides of the Plexiglas Pyramidion.

STEP 5:

After the Paint dries, remove the Painter's tape, insert the interior light, and attach the Plexiglas Pyramidion capstone to the Pyramid structure.

HOW TO MAKE AN ORGONE PYRAMIDION CAPSTONE

The term **"Orgone"** refers to a "universal life force energy" which is suggested to be harnessed and accessed by using "Orgone Pyramids". While the concept of "Orgone Energy" is considered pseudoscientific, the construction method of "Orgone Pyramids" can serve a practical use as a Pyramidion or "Pyramid capstone". The following are instructions on how to make an Orgone Pyramidion .

Ingredients:

* Polyester resin kit
* Rubber gloves
* Cooking oil
* Plastic containers
* Paint stir stick
* Paper towels
* Metal shavings
* Bamboo skewers
* Copper wire
* Quartz crystals
* Sharp Knife
* Ruler
* Tape, Glue
* Nails
* Protractor
* metal, copper, tin, aluminum, or wood to build the mold

STEP 1:

Use metal, copper, tin, aluminum, or wood to build the mold for the Orgone Pyramid Capstone. If your pyramid was designed at any of the angles of the Polygon shapes shown on page 167, then you can use these polygon shapes as a template to build the Orgone Pyramid capstone mold. For the Squared-Circle Pyramid designed with a 72° degree angle, we can use the Decagon shape.

STEP 2:

Draw the polygon shape onto the mold material where the length of each side of the shape will be the base of the capstone, and use a sharp knife to cut off the unneeded portions

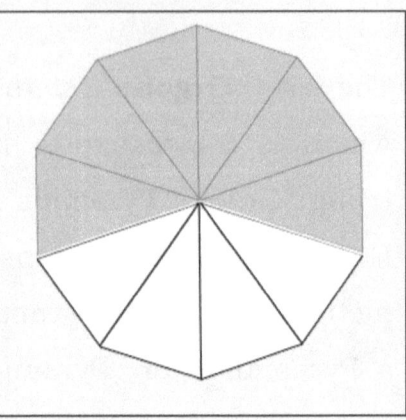

of the shape keeping only 4 triangles.

STEP 3:

Use a sharp knife to score the 3 lines of the 4 remaining triangles. After scoring, fold the 4 triangles together and seal the open side with tape, glue, or nails depending on the material being used. The finished result will be a small Pyramidion mold shape with an open bottom in which the Orgone material can be poured.

STEP 4:

Spray or cover the inside of the Orgone Pyramidion mold that you just constructed with Oil or Lubricant.

STEP 5:

Turn the Orgone Pyramidion Mold upside-down and place it in a box where it can remain supported and stable while upside-down.

STEP 6:

Put on your rubber gloves and pour the Polyester resin powder into a plastic container where the quantity is enough to fill the Pyramidion Mold.

STEP 7:

Add the liquid activator from the Polyester resin Kit to the Powder in the plastic container and stir the mixture using a paint stir stick.

STEP 8:

Put various decorative materials inside the Orgone Pyramidion mold prior to filling it with the Polyester resin. Some of the popular decorative materials include various crystals, metallic shavings, and copper wire coils.

STEP 9:

Pour the mixed Polyester resin into the Orgone Pyramidion mold being sure to cover all of the decorative objects and filling in the spaces between the objects placed inside of the mold. The process of adding decorative objects and pouring in the Polyester resin can be done in layers inside the Pyramidion mold to give the finished capstone a different appearance.

STEP 10:

Place the Orgone Pyramidion mold which contains the decorative objects and the Polyester resin in the Sun or in a warm dry place for 8 to 10 hours to allow the resin to harden and cure.

STEP 11:

Once the Polyester resin has hardened inside of the Orgone Pyramidion Mold, turn the Pyramidion over and the Orgone Pyramidion capstone will slide out of the mold. The capstone can then be fixed and attached to the Pyramid as a decorative finishing piece.

OBOADEE

7.0. CHRONICLES OF A PYRAMID BUILDER

In 2011, the craft of Pyramid Building was taken up by a descendant of the Ancient Pyramid Builders of Africa. These are the Chronicles of Osiadan BoreBore Oboadee and his journey into the Ancient African Architectural Art of Pyramid Building.

February 22, 2011

I have decided that I would like to attempt to build a Pyramid. I have been noticing that the Geometry of a Pyramid can be approximated by criss-crossing two parabolas; this point is most evident in the "Bent Pyramid" found in Egypt. I also have been doing research into an Architectural structure called "Nubian Vaults" and "Nubian Domes" where it is apparent that our Ancient ancestors in Africa had knowledge of building parabolic Arches, Vaults, and Domes. Maybe I can criss-cross two Nubian-Vaults and come up with a Pyramid-like structure. I will build this structure in my back yard here at my home. I have noted that my home is on the 33.7° N Latitude marker and the Giza Pyramids are on the 30° N Latitude marker; a difference of only 3.7° which may be significant.

March 5, 2011

I completed the design for the "Dual Arch Pyramid". I decided on using a poured concrete square slab as the level foundation. The Arches will be 108 inches wide and 75 inches tall on the interior. I measured and marked off the area needed in my back yard. The final structure may not look exactly like a Pyramid, but once completed I can add some other bricks to the structure to give it more of a Pyramid shape.

March 17, 2011

I bought the building Materials for the Dual Arch Pyramid from the hardware store today. The total bill was about $165. The items I bought were:

- 5-80 pound bags of concrete @ $5.00 each = $25.00

- 129 common red bricks @ $0.50 each = $64.50

- 4-50 pound bags of mortar @ $4.00 each = $16.00

- 17-2'×4' wood planks @ $3.00 each = $51.00

- 2-1'×12' plywood planks @ $4.00 each = $8.00

I'll start working on the project this weekend. Hopefully it turns out good!

March 19, 2011 – Marcy 20, 2011

This weekend, I broke ground on the Dual Arch Pyramid project. The area I am building on is slanted, so I had to do some ground Grading and leveling. On Sunday, I laid the concrete for the foundation. I didn't know how difficult it was to mix concrete by hand.

Above: Leveled, graded area with wood frame before pouring concrete

March 21, 2011 – March 25, 2011

This week I built the formwork for the brick Arches. Part of the plywood broke on one side of the wood arch when I was bending it across the wood arms so I used a metal sheet to fill in the missing gap.

Above: wood formwork for the Brick Arch

March 26, 2011 – March 27, 2011

Today I began laying the bricks for the Arch. I ran into some trouble when the wood formwork that I patched with the metal sheet started to bend under the weight of the Bricks. I added some more support arms to the form to give it more stability. However, the bend in the Brick Arch is obvious, and I am afraid the Arch will collapse when I remove the wood formwork.

Above: First Brick Arch laid across the wood formwork

April 1, 2011 – April 3, 2011

This weekend I am still letting the brick arch dry and set. I am doubtful that when I remove the wood formwork that the brick arch will still be standing. I started experimenting with trying to build a regular square pyramid by offsetting each layer of bricks by corbelling, but the experiment collapsed after just four layers. However, I tested another method, but instead of offsetting successive square layers, I tried offsetting successive circular layers making a cone, and I was able to keep build up until I ran out of bricks. I think if I can combine the two methods, then I can build a true Pyramid structure.

Successive offset square layers of bricks collapsed after just 5 layers

April 8, 2011 – April 10, 2011

I have labeled this weekend "the moment of truth" because this is the weekend when I have to remove the wood formwork from underneath the brick arch and see if the brick arch will still stand. To my surprise, I was able to remove the wood formwork and the brick arch stayed up even with the misshapen section. I am now optimistic with going forward with this project.

Above: First Brick Arch after removing wood formwork

April 15, 2011 – April 17, 2011

I rotated and positioned the wood formwork to lay the bricks for the second Arch. Also, during the week, I worked out the design for a second Pyramid building project that I would like to try that should result in a true Pyramid form. I am calling this second project, the "Squared-Circle" Pyramid because I had to combine the geometry of the Circle and the Square in order to come up with a design that would not collapse and could be scaled-up to build larger structures.

Rotated and Positioned the wood formwork to lay the bricks for the second Arch

April 23, 2011 – April 25, 2011

This weekend I laid the bricks for the second Arch. It's funny, even though I successfully removed the wood formwork from the first Arch, after I finished laying the bricks for this second Arch, I started feeling the same uneasy feeling that the Arch would not stay up. Well, I will see once it's time to remove the wood.

Finished laying brick Arch across wood formwork

April 30, 2011

This weekend I am still waiting for the second brick arch to dry and set. I went to the Hardware store and bought the building Materials for the "Squared-Circle Pyramid". The total bill was about $212. The items I bought were:

- 330 common red bricks @ $0.50 each = $165.00

- 5-50 pound bags of mortar @ $4.00 each = $20.00

- 2-12"x18" Plexiglas panes @ $5.00 each = $10.00

- Plexiglas cutting knife @ $5.00

- Silicone Sealant @ $5.00

- Plastic Epoxy @ $4.00

- Water proofing tape @ $3.00

I'll start working on the "Squared-Circle Pyramid" project the weekend after I remove the wood formwork from underneath the second brick arch.

May 14, 2011

This weekend I removed the wood formwork from underneath the second Brick Arch. It's good to have completed this project and I look forward to the next challenge.

Finished Binary Brick Arch Pyramid with formwork removed

June 04, 2011 – June 05, 2011

Today I Broke Ground to lay foundation for Squared-Circle Pyramid. After mixing the cement for the concrete slab for the foundation of the Dual-Arch Pyramid, I decided to only lay concrete for the foundation of the Squared-Circle Pyramid in the area where I will be laying the bricks. So the interior of the foundation layer is still dirt, and the foundation layer is a Concrete Square slightly wider than a brick.

Preparing Foundation layer of the Squared-Circle Pyramid

June 08, 2011

Today I completed the First Layer of Bricks for the "Squared-Circle" Pyramid. Laying the bricks for the outer square is not difficult; however, using the straight-sided bricks to form the inner circle is more challenging. I have found that cutting a brick into 4 equal triangles helps me to better approximate the inner brick circle. Because this structure is being built around a light, I had to leave a space between the bricks for the light's cord. To save money on materials, instead of filling in the corners between the brick circle and square with concrete, I filled the layers with compressed soil.

Layer 1 of the Squared-Circle
"Pyramid of the Midnight Sun"

June 18, 2011

Today I completed the Second Layer of Bricks for the "Squared-Circle" Pyramid. I was able to cover the space between the bricks that I left for the light's cord and bury the exposed part of the light's cord after completing this layer.

Layer 2 of the Squared-Circle
"Pyramid of the Midnight Sun"

June 25, 2011

Today I completed the Third Layer of Bricks for the "Squared-Circle" Pyramid.

Layer 3 of the Squared-Circle
"Pyramid of the Midnight Sun"

June 29, 2011

I completed the Fourth Layer of Bricks for the "Pyramid of the Midnight Sun" project today. Because I am using triangular shaped bricks to approximate the inner circle, the interior is starting to look like an octagon. This is making me want to build an "octagon" shaped Pyramid. Maybe one day in the future. I know that the headquarters for the American Institute for Architects is an Octagon shaped building, and the nickname for the octagon is the "squared-circle", so perhaps this is significant.

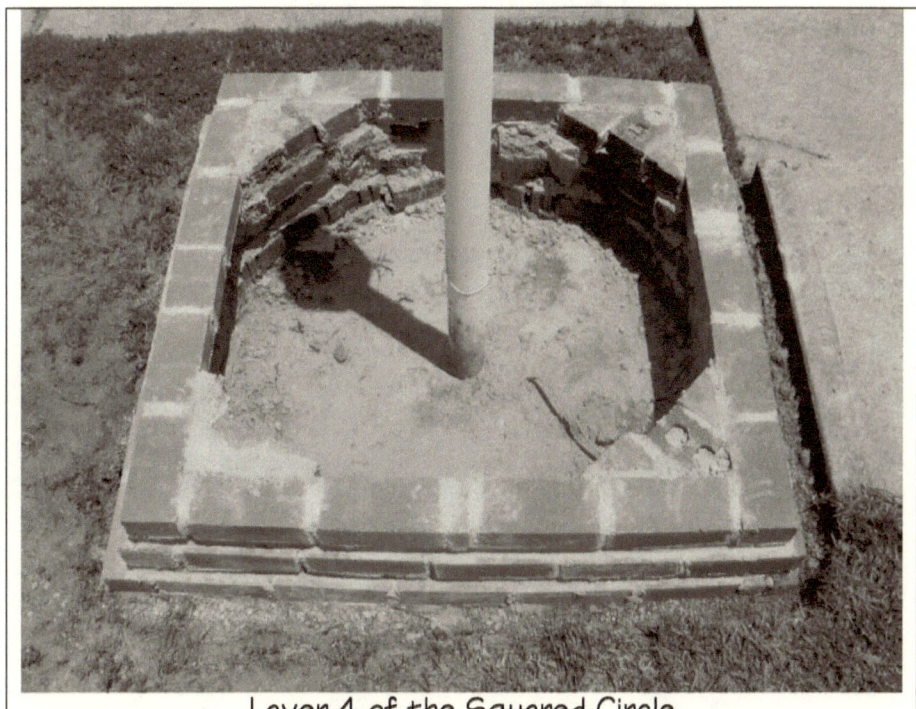

Layer 4 of the Squared-Circle
"Pyramid of the Midnight Sun"

July 02, 2011

I finished the Fifth Layer of Bricks today. Everything is coming along. The structure is still too short to really look like a Pyramid yet.

Layer 5 of the Squared-Circle
"Pyramid of the Midnight Sun"

July 04, 2011

I finished the Sixth Layer of Bricks today. I thought about painting the structure once I am finished, but I think I like the look of the raw brick masonry better.

Layer 6 of the Squared-Circle
"Pyramid of the Midnight Sun"

July 06, 2011

I completed the Seventh Layer of Bricks. I guess at this point the structure qualifies as a small trapezoidal mastaba. I'll keep on building up

Layer 7 of the Squared-Circle
"Pyramid of the Midnight Sun"

July 08, 2011

Eighth Layers of Bricks done. I am noticing that it gets easier as I go up. I have only been doing 1 layer a day for the time that I have set aside for this project, but as I get towards the top, I will be able to do multiple layers a day.

Layer 8 of the Squared-Circle
"Pyramid of the Midnight Sun"

July 10, 2011

Today I finished the Ninth Layer of Bricks. I have been neglecting my yard work to work on this project. When I am done, I am going to start thinking of some landscaping ideas to complement the Pyramid I am building. Maybe if I can find a Lion Sphinx stature for the yard that would be a nice complement.

Layer 9 of the Squared-Circle
"Pyramid of the Midnight Sun"

July 12, 2011

Ten layers of bricks completed for the "Squared-Circle" Pyramid.

Layer 10 of the Squared-Circle
"Pyramid of the Midnight Sun"

July 14, 2011

Today I finished the Eleventh Layer of Bricks for the Pyramid.

Layer 11 of the Squared-Circle
"Pyramid of the Midnight Sun"

July 21, 2011

I Completed layers 12, 13, and 14 of the "Squared-Circle" Pyramid today. This was the first time I completed multiple layers in a day for this project. The structure is starting to look like a Pyramid.

14 Layers completed of the Squared-Circle
"Pyramid of the Midnight Sun"

July 25, 2011

Today I Completed layers 15, 16, 17, and 18 of the "Squared-Circle" Pyramid of the Midnight Sun project. Everything is looking good, I am just about finished, I just have 1 more layer to go and then I have to do the capstone and I will be done.

18 Layers of the Squared-Circle "Pyramid of the Midnight Sun"

July 27, 2011

Today I Completed layer 19, and I also Removed Post Light so that I can place a Pyramid shaped light cover as the capstone

19 Layers of the Squared-Circle
"Pyramid of the Midnight Sun"

July 28, 2011 – August 05, 2011

I Began working on the Capstone Light for the Pyramid of the Midnight Sun. I also glued small mirrors to the interior base of the Pyramid to reflect the light.

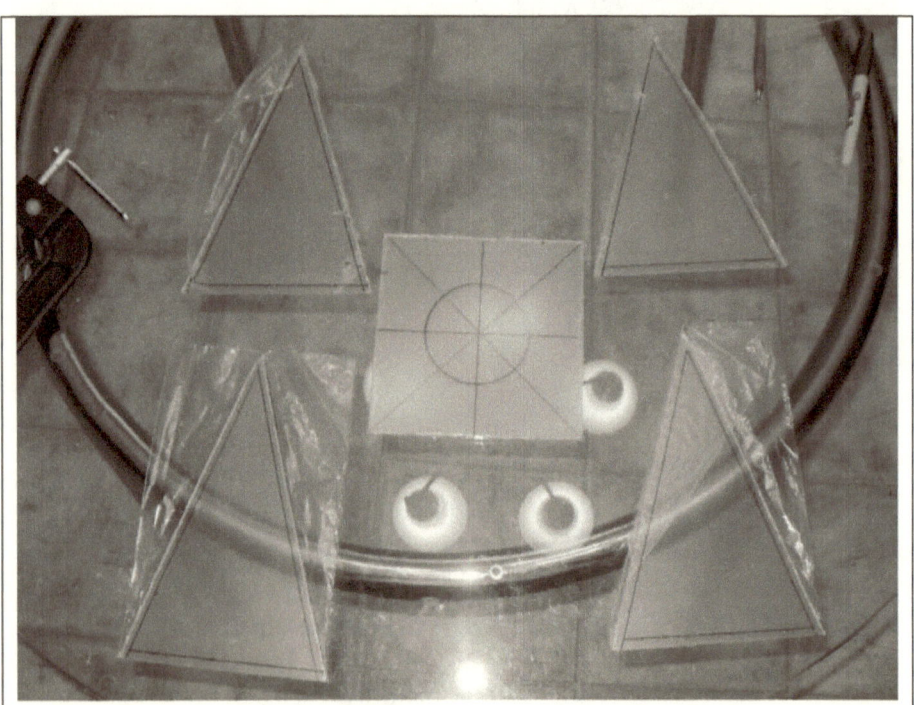

Cut out 4 Plexiglas Triangles and 1 Plexiglas square for the Capstone Light

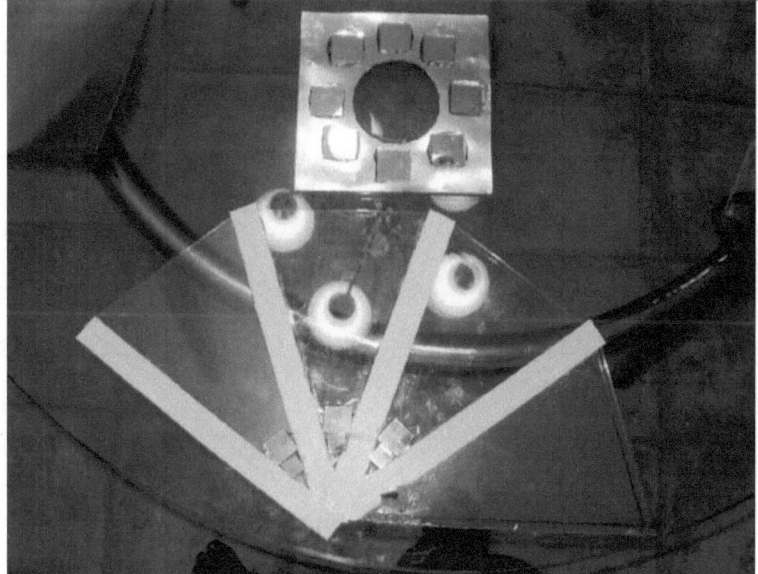

Above: Combining the Triangles to form the Plexiglas Pyramidion Capstone for the Squared-Circle Pyramid

Above: Finished BenBen Pyramidion Capstone light cover for the Squared-Circle Pyramid

August 23, 2011

Today I placed the Capstone on the "Squared-Circle" Pyramid and watched the "Pyramid of the Midnight Sun" Light at Sunset for the first time. I am satisfied with the results and really look forward to scaling this project up and training others to build Pyramids so that we can build massive Pyramids like the Ancients all over Africa

Squared-Circle
"Pyramid of the Midnight Sun"
with Lit Capstone

8.0. ABOUT THE ARCHITECT & AUTHOR

Greetings, How are you? Who Am I, Why I am **Osiadan Borebore Oboadee**, the Author and Architect of this great work (**magnum opus**) entitled "**ARCH I TECT: How to Build A Pyramid**". The reader may want to know Who (**Hu**) I Am, but what is more important is the fact that the information I present is accurate, practical, and useful. Furthermore, the reader can confirm the accuracy of the information I present by applying the concepts and seeing that what I present does indeed work in theory as well as application!

I was born and raised in the Western nation known as the United States of America; however, my lineage can be traced back to the Ancient Neolithic Pre-Dynastic Architects and Builders of Africa. I am a descendant of the **Balanta-Bassa** and **Djola-Ajamatu** tribes in present day **Guinea-Bissau** (**Ghana-Bassa**) West Africa. Both the Balanta and Djola tribes migrated in Ancient times from the area which is present day **Egypt**, **Sudan**, and **Ethiopia**. I am a Scientist, Engineer, Mathematician, Problem Solver, Analyst, Synthesizer, Artist, Craftsman, and Technologist by education, profession, and nature.

My formal education and training has led me to obtain Bachelors and Masters Degrees in the areas of Electrical Engineering, Physics, and Mathematics between the years of 2003 and 2006. I have a keen interest in Architectural design and I have practiced masonry as a hobby, but I have had to rely on the autodidactic ability of being self-educated in the areas of Architecture and construction. While I have worked as an Electrical Engineer and Computer Programmer for several years, I have always had a strong interest in Ancient and Traditional African Culture and Philosophy. It was my interest in African Culture and Philosophy that motivated and inspired this book. I made my first visit to Ghana, West Africa in 2008. After my visit to Ghana, I adopted the title **"Osiadan Borebore Oboadee"** which comes from the Twi language spoken in Ghana, West Africa.

The Twi word **"Osiadan"** comes from the root words "Si" meaning "Build" and "adan" meaning "Building" with "O-" being a way to denote a "Master", hence "Osiadan" literally describes a **"Master Builder of Buildings"**. Also note the phonetic similarities between the Twi words "Si" and "Adan" and the Ancient Egyptian words **"Sia"** (wisdom) and **"Aton"** (high noon sun). The Twi word **"Borebore"** comes from the root words "Bo" meaning "Create" and "Re" meaning "to do repetitiously", thus

"Borebore" is used to describe a **"Continuous Creation"** or **"Architect"**. The word "BoreBore" or "Bore" in Twi is also related to the Hebrew word **"Bara"** meaning **"to begin"** found in the first verse of the first chapter of the Judeo-Christian Bible, and to the Yoruba word "bere" meaning "to begin". Also note the phonetic similarities between the Twi words "Bo" and "Re" and the Ancient Egyptian words "Ba" (soul) and "Re" (sun). The Twi word **"Oboadee"** comes from the root words "Bo" meaning "Create" and "Abode" meaning "Creation" with "O-" being a way to denote a "Master", hence "Oboadee" literally describes a **"Master Creator of Creations"**. Oboadee is also pronounced O-Poatee in different African dialects, and is said to derive from the pronunciation of the name of the Ancient African Creation deity PTAH. Osiadan, Borebore, and Oboadee are three principles of **African Creation Energy** used as attributes of "God" in various African cosmologies.

Between the years of 2009 and 2010, I wrote a three books dedicated to the Creative Liberation of African people entitled **"The African Liberation Science, Mathematics, and Technology (S.M.A.T.) Project"**. The titles of the three books that are a part of the "The African Liberation S.M.A.T. Project" are:

1. The SCIENCE of Sciences and The SCIENCE in Sciences

2. 9^{9^9} Supreme Mathematic, African Ma'at Magic

3. P.T.A.H. Technology: Engineering Applications of African Sciences

The Triad of Books part of African Creation Energy's "African Liberation Science, Math, and Technology Project": 1 - "The Science of Sciences and The Science in Sciences", 2 - "Supreme Mathematic, African Ma'at Magic", and 3 - "P.T.A.H. Technology: Engineering Applications of African Sciences"

My primary purpose for writing the books in the "African Liberation S.M.A.T. Project" was to motivate the Creative Energies, Minds, and Bodies of African people to go from an inert state of Theory and Speculation to an Active creative state of Development, Creation, and Productivity for the survival and well-being African people everywhere. Now in 2011, at the age of 30, after writing the books of the "African Liberation S.M.A.T. Project", following the Ancient African Tradition, I found it necessary to provide evidence of the Philosophy in Action and Application by building structures. Thus, I embarked upon this project. Moreover, In Ancient Nile Valley culture, a celebration occurred every **30 years** called the **"Djed Festival"** which represented the resurrection, rejuvenation, and renewal. I was born in the year **1981**, and the construction of the pyramids for this project, and the authoring of this text in the year 2011 is how I, **Osiadan BoreBore Oboadee**, have symbolically chosen to celebrate my **30 year "Djed Festival"** of renewal for all eyes to see.

Following the multitude and plethora of information presented by the many Scholars (who we have affectionately labeled **"MASTER TEACHERS"**) who have came to improve the conditions of African people namely: The **Honorable Marcus Mosiah Garvey, Noble Drew Ali, the Honorable Elijah Muhammad, Clarence 13X "Father Allah", El-Hajj Malik El-Shabazz "Malcolm X", Dr. Cheikh Anta Diop, Dr. John Henrik Clarke, Dr. Yosef Ben-Jochannan, Dr. Théophile Obenga, Ivan van Sertima, Mathu Ater, Ashra Kwesi, Anthony Browder, Asa Hilliard, Patrice Lumumba, Kwame Nkruma, Amílcar Cabral, Ra un Nefer Amen, Ben Ammi, Dr. Malachi York "Amun Nubi Re Ah Ptah"** and many others, it is the goal of **Osiadan BoreBore Oboadee** and **African Creation Energy** to be the catalyst in the synthesis, unification, and practical application of all of the information presented by the great Master Teachers. Thus, it is the aspiration of Osiadan Borebore Oboadee and "African Creation Energy" to be a **"Master Technician"** and Te<u>Ach</u> through Action and Application. In keeping with this goal, the great work of Training other "Master Technicians" has begun. Using **African Creation Energy**, **Osiadan Borebore Oboadee** has initiated the activation of the "Master Technicians" around the world in the practical application of the information presented by the various "Master Teachers". Training Master Technicians using African Creation Energy will provide the Architects, Engineers, and designers of the future African Industrial Infrastructure needed to develop Africa to the point where it is a viable and desirable destination for the mass repatriation of the African Diaspora. The Architecture which will create the environment that will program the minds of the future Master Technicians of Nature must itself be in tune with Nature. Architecture that is in tune with Nature and "Ecology" has been given the name **Arcology**, and Arcologist strive to design Architectural hyper-structures that create environments that are self-contained and self-sufficient. Examples of some Arcology structures include the Shimizu Mega-City Pyramid, the New Orleans Arcology Hyperstructure (N.O.A.H.) Tetrahedron, and the Dubai Ziggurat-Pyramid. It is paramount that African Creation Energy be applied in the design and construction of African Arcological Nature Architecture in order to sustain Liberation and Self-Sufficiency into the future for African people. It has been my honor to present this work which I intend to be a cornerstone and building block in the construction of our future African Nature Architecture.

I'M SATISFIED

I'M at PEACE

I'M HOTEP

Son of PTAH

Kwadwo Pado
Osiadan Borebore Oboadee
"Prophessor A.C.E."
www.AfricanCreationEnergy.com

REFERENCES AND RESOURCES

* "A Home Fit For Royalty: How to Build an Economically, Ecologically, and Indestructibly Efficient Home" by Jarred Wesley

* "Abu Simbel to Ghizeh: A Guide Book and Manual" by Yosef Ben-Jochannan

* "Africa: Mother of Western Civilization" by Yosef Ben-Jochannan

* "Architecture and Mathematics in Ancient Egypt" by Corinna Rossi

* Ari-Kat: 1-The Revelation of the Pyramid Stones, 2-The Alchemist who Built the Pyramids by Relevant Televison http://www.relevant-television.com

* Auroville Earth Institute http://www.earth-auroville.com

* "Black Genesis" by Robert Buval and Thomas Brophy, Ph. D

* "Building with Earth: Design and Technology of a Sustainable Architecture", Second and Revised edition by Gernot Minke

* Geopolymer Institute http://www.geopolymer.org

* History of ancient Egypt, Volume 1 By George Rawlinson

* Hurry, Jamieson Boyd, Imhotep, the Vizier and Physician of King Zoser, and Afterwards the Egyptian God of Medicine, AMS Press, 1928, 1978.

* "Mathematics in the Time of the Pharaohs" by Richard Gillings

* Mesopotamia – The British Museum http://www.mesopotamia.co.uk

* Pyramid Texts Online http://www.pyramidtextsonline.com/translation.html

* "Temples Of The African Gods" by Michael Tellinger

* "The Complete Guide to Masonry & Stonework", Created by: The Editors of Creative Publishing International, Inc. in cooperation with Black & Decker.

* "The Pyramid Builders of Ancient Egypt: A Modern Investigation of Pharaoh's Workforce" by Rosalie David

* "The Pyramids of Egypt" by I.E.S Edwards

* "The Secret Diaries of Hemiunu, Architect of the Great Pyramid" Translated and Interpreted by Derek Hitchins

www.ingramcontent.com/pod-product-compliance
Lightning Source LLC
Chambersburg PA
CBHW031836170526

45157CB00001B/326